Lionel Chalmers

An Essay on Fevers

More particularly those of the common continued and inflammatory kinds

: wherein a new and successful method is proposed for removing them

speedily : to which is added, an essay on the crises of those disorders

Lionel Chalmers

An Essay on Fevers
*More particularly those of the common continued and inflammatory kinds :
wherein a new and successful method is proposed for removing them speedily : to
which is added, an essay on the crises of those disorders*

ISBN/EAN: 9783337383961

Printed in Europe, USA, Canada, Australia, Japan

Cover: Foto ©berggeist007 / pixelio.de

More available books at **www.hansebooks.com**

AN
ESSAY
ON
FEVERS;

MORE PARTICULARLY THOSE OF THE

COMMON CONTINUED AND
INFLAMMATORY KINDS:

WHEREIN

A NEW AND SUCCESSFUL METHOD IS
PROPOSED FOR REMOVING THEM
SPEEDILY.

TO WHICH IS ADDED,

An Essay on the Crises of thofe Diforders.

By Lionel Chalmers, M.D. of Charles-Town.

———— *Fungar vice cotis.* ———— Hor.

LONDON:
Printed for Edward and Charles Dilly, in the Poultry.
M.DCC.LXVIII.

ADVERTISEMENT.

DR. JOHN LINING having profecuted a courfe of ftatical experiments, for the fpace of one year at *Charles-Town*, the author of the following ESSAY, became, afterwards, fo clofely connected with him, both by mutual intereft and friendfhip, as gave him an opportunity to be fully informed of all he defired to know, relating to the perfpiration and other natural difcharges in each feafon : By which means, he hath been led to judge otherwife of them than *Sanctorius* did. It is difficult to conceive on what grounds the latter founded his notions, with refpect to the numerous ill confequences he imputes to an abatement of the perfpiration alone : nor is it lefs furprifing, that fo many great men fhould have fo implicitly adopted his miftakes.

The author hath fuppofed, that a fpafmodic conftriction of the arteries and other mufcular membranes, is the immediate caufe of fevers, in whatever way this may be brought on. For all the fymptoms in acute difeafes fhow fomething of this fort; the feverity or

mildnefs

mildnefs of thofe diftempers, for the moft part, keeping pace with the degree of that conftriction, and the number of veffels it affects.——If this term fhould difpleafe, another of the fame meaning may be ufed in its ftead.

Having oftentimes been difappointed in his practice, though he followed the methods recommended by the beft writers, fo far as they were underftood by him, he undertook to keep a journal of the weather, and obferve the difeafes that accompanied its changes, in the feveral feafons. Which he did for many years in hopes of improvement thereby; examining his patients with the greateft exactnefs at the fame time, fo as to note every favourable and unfavourable circumftance through the whole progrefs of the difeafe. This foon difcovered, that when certain fymptoms occurred at the beginning of a fever, the complaint, generally, was mild and of fhort continuance; nor would the patient be feized with any inflammatory difeafe, though fuch did not fail to attack others, who fickened at the fame time, but without the like favourable incidents.——After carefully attending to this in great numbers of cafes, the hint was

not

not flighted; for what happened of its own accord, and prevented fuch diftempers, was brought on by art to remove them, after they were fixed.

Nothing was more wanted, than a fafe, fpeedy, and, as far as the nature of things will permit, a certain method to remove fevers, within the firft and fecond day. This furely would well deferve the name of a cure: But when they are allowed to run through the feveral ftages, into which they would pafs were they left to their own courfes, and the patients recover at laft, all the merit that then can be pleaded is, that nothing improper had been done. Happy would it be could this be faid on all occafions.

It may, perhaps, be faid, that though inflammatory and other common fevers may probably yield to the following treatment in warm countries, yet they will fcarcely do fo in cold. The infinuation is more fpecious than true; for the human body is of the fame conftruction every where, differing only in being more braced in one climate, and laxer in another: And, truly, the inhabitants of *South - Carolina* vary as much in thofe refpects

fpects, during the winter and fummer, as if they dwelt in climates the moft oppofite in temperature. Befides, the fame diftemper will be much alike in its fymptoms and events, and, therefore, muft require the fame management, or nearly fo, in all places.

The author hath practifed in the way hereafter laid down, for many years, and with fuch fuccefs, that he is under no apprehenfions of its failing often, provided his directions be but duly followed. And having dealt candidly in publifhing it, in hopes it may be of fome ufe to the people of this province, candour requires that it fhould not be judged of by the fallacious rules either of prepofleffion or prejudice, but fair experiments only. He warns the reader, however, that neither this nor any other method will at all times cure; for every one of the profeffion muft have found in the courfe of his practice, that fomewhat of a hidden nature now and then occurs, even in the complaints he thought himfelf beft acquainted with, which, in fpite of his boafted fagacity, difappointed him. Whether this depended on certain fingularities in the conftitutions of

4   individual

individual perſons, the nature of the diſtem-
per itſelf, or his own overſights, it might not
be poſſible for him to diſcover ; or even if he
could, it might not then have been in his
power to remedy it.

It was deemed ſufficient to point out the
method, and mention the medicines to be
uſed, without determining their proportions,
or when it may be proper, either to omit, or
prefer any one of them to another ; for every
practitioner ought to know, how they ſhould
be adjuſted to the different ages and conſtitu-
tions of his patients.

At the end of this Essay is *another* on the
Criſes of Fevers, which is intended to confirm
the reaſoning contained in the former ; by
ſhowing that a criſis could not happen, unleſs
the fluids had firſt been too much accumulated, in or near the places whence the diſcharges called critical proceed, in conſequence
of ſome hinderance to a free circulation in
other parts.

CONTENTS.

ESSAY

E S S A Y I.

C H A P. I.

An enquiry whether an abatement of the perspi-
ration be, indeed, the cause of so many di-
stempers as are commonly imputed to it.

HOUGH *Sanctorius* proved that
our bodies perspire freely, he seem-
ed to have no knowledge of their
imbibing any thing from without
except air ; and therefore was not
aware, that his experiments discovered no
more than the difference between the perspiration and moisture we absorb through
the pores of the skin and lungs: but exhala-
tion and absorption so confound each other
as to make it extremely difficult, if not im-
possible, to ascertain the true sum of either.
We may be assured, however, that the dis-
B charges

charges from the lungs and external furface are greater than fuch trials fhow them to be ; fince no allowance is made for whatever we imbibe, which, beyond all doubt, muſt be very confiderable (1). When, therefore, that author found his weight increafed, he always charged it to fo much of the perfpiration being detained; and afcribed to fuch accidents almoſt every difeafe to which mankind is liable. But this was, by no means, a juſt inference: for it cannot be thought that he experienced all thofe ill confequences in himfelf; and to apply the refult of experiments, which concerned him alone, to whatever happened to others, in whom that, as well as all the other difcharges, might have proceeded in a very different manner, was ſtretching the conclufion rather too far. Indeed, we owe much to *Sanctorius*; but, ſtill, there is no need to admit his miſtakes: for it is in part owing to this condefcenfion, that I know not

(1) A negro man who had eaten or drank but little for fome days before he was gibbetted at *Charleſtown* in *March* 1759, and had nothing given him afterwards, regularly voided a large quantity of urine early every morning; but he difcharged no more till about the fame hour next day. The reafon undoubtedly was, that fo long as the fun was above the horizon and warmed the air, perfpiration was freer, becaufe the fluids were more determined towards the fkin thereby; but as it became colder and more damp in the night, the outward furface was then conſtricted, and the humours, confequently, being repelled towards the vifcera, urine was increafed, the moiſture abforbed furnifhing matter for it.

what

what attenuants and fudorificks have been introduced into the practice, in order to diffolve some imaginary vifcidity of the fluids, and promote fweats in fevers; on which two effects, the removal of moft acute difeafes is thought chiefly to depend, though one of them acts in direct contradiction to the other.

Some men of great reputation have likewife adopted *Sanctorius*'s affertion, that five-eighths of all our food and drink ought to pafs off by perfpiration : but this rule cannot hold in any country, where cold and warm feafons fucceed each other, whatever may happen at times during the Summer. Nor is it lefs an error to maintain, that the daily fum of all the difcharges fhould equal the quantity of our diet and drink, to keep up the fame weight of body that, we may enjoy health. For to anfwer that great end, thofe excretions muft vary at different times of the year; and therefore, no fixed weight can be affigned for the ftandard of health : becaufe that which would be moft proper at one feafon, might be dangerous in the higheft degree at another; as will appear in the courfe of this effay. It feems, indeed, that this great perfpiration was meant as neceffary, only in the Summer. For from the autumnal equinox to the Wnter folftice, he himfelf informs us, that it fcarcely exceeded one pound, cr was about one fifth-part of what it had been at another time : yet

he

he does not complain of any diforders from fo
remarkable a difference therein : as certainly
this abatement happened by degrees ; and in
all likelihood it continued thus throughout the
month of January, and for a good part of February. In general, as the weather is cooler and
more changeable in the Spring and Autumn
than in Summer, perfpiration not only muft
be lefs then, but more variable alfo. How
dangerous therefore, or at leaft troublefome
and difficult muft it be, to follow his rule for
the prefervation of health, by reducing the
body in the Spring to the weight it fhould be
of in Summer, and keeping it to that poife.
But this reafoning would equally hold for pre-
ferving the like weight of body in Winter, as
during a warm feafon, by promoting perfpiration to the fame quantity in the former period that it ought to be in the latter. He,
moreover, fays, that they will enjoy the beft
health, and may attain extreme age, who keep
their bodies from varying much in weight.
Under what limitations *much* is to be underftood here, I know not : but we are well affured, that the perfpiration ought to increafe
with the heat of the air ; elfe we could not
live, much lefs enjoy health, during very
warm weather (1). For as the fiery particles

(1) Dr. *Lining* began to fall away in March, and by
the end of September loft feventeen pounds weight : but
he regained that and more, before the end of February
following.

enter

enter our bodies along with the watry ones, and, perhaps, some portion of the air itself, unless the fluids were at the same time constantly discharged, according to the expansions they are made to undergo, and also to free us of this unnatural heat, such a general overfulness must soon ensue, as would distend the vessels beyond all power of contracting themselves: instances of which happen almost every summer to some people in *Carolina*. This profuse perspiration is therefore absolutely necessary, during a sultry season, to preserve a balance between the much relaxed vessels and highly rarified fluids. Hence it appears, that our bodies are constantly discharging the heat they absorb from without: and so languid do the vital actions become, that whilst we sweat freely as we ought to do, our natural heat is really less after the Summer is far advanced than in Winter. In January 1752, when the mercury fell to the 18th division, it rose in a small thermometer placed under my arm, to the 97th : but in July that same year, when the heat of the shaded air was marked by the 101ft degree, it did not rise in the same situation above the 95th.

It was just now said, that we could not live when the weather is very warm, unless the perspiration increased in proportion as the fluids were rarified. Accordingly, I sometimes have been sent for to people who were expiring, in

a manner,

a manner, from being too long expofed to the intenfe heat of the fun, and to others who were already dead before I could reach them, though but a few doors from me. The firft fymptoms perceived on thefe occafions, are an univerfal languor and a faintifh fort of drowfinefs: a tightnefs is next felt round the breaft, as if it were ftrongly girded, fo that the perfon can fcarcely breathe: and foon after he falls, as it were, into a profound fleep; from which if he cannot be fpeedily roufed, death quickly follows. In fuch cafes, I commonly found the fkin rather dry and very hot; the countenance full and flufhed; and all the outward veins very much diftended. On the breaft, and other places, inflammation might be perceived, as in the fcarlet fever, or a flight eryfipelas, the whole body appearing enlarged and fuller than in health; but the rednefs we fpeak of in the fkin, only happened a little while before the patient's death. The putrefaction in thofe who die after this manner, is always fudden and great; which makes it neceffary to bury them in a few hours: nor does the heat their bodies acquired abate fo much after death as in other carcaffes, for the time they lie uninterred. But thefe things by the bye——

Dr. *Lining* (who was very exact) fometimes found himfelf heavier by feveral ounces, than he was a few hours before, though he
perfpired

perfpired fenfibly all the while from feveral parts of his body, as well as from the lungs : hence it feems, that perfpiration and abforption may both go on freely together. But he oftentimes abated the perfpiration defignedly, without any perceiveable injury, fo that the matter detained, either was not fo pernicious as is commonly thought, or his body muft have been quickly freed of it by fome other difcharge. So earneft was he to know what might be the confequence of checking this fuddenly, that he fometimes expofed himfelf at once to the wind, when fweating profufely. But this curiofity was not always gratified without difagreeable effects. In particular, at the beginning of July, having only an open thin linen waiftcoat over his fhirt, he fat from eleven till noon in a clofe room where the heat was marked by the 87th degree of *Fahrenheit*'s thermometer ; and in that one hour perfpiring eleven ounces and an half, he inftantly removed into an open paffage through which the wind blew brifkly, the mercury falling there to 82 : by which, in one hour and a quarter, he gained fix ounces and an half, though he ftill fweated from fome parts. The whole of this muft furely be charged to abforption, and confifted either of the perfpiration with which his linen was damped, or elementary water imbibed from the air, not more than 13 or 14 degrees colder than his blood, which, of itfelf, was not likely to do

B 4   much

much harm. However, let us even grant
that fome of the perfpiration was detained
here, yet the quantity thereof, together with
whatever elfe was abforbed, did not occafion
any great fulnefs or plethora ; nor could a putrefcence enfue immediately. But fuch experiments were unfafe on another fcore; and
this we fpeak of, was prefently followed by
gripings, and a painful wearinefs chiefly affecting the legs, knees and thighs : however
thofe fymptoms went off on his having fome
plentiful ftools.—Were not thofe effects owing
to the fudden ftrong conftrictions which the
outward veffels and membranes were made
to undergo, by the cooler air, and the hindrance thence arifing to the motion of the
blood through thofe parts, from which it was
repelled and made to accumulate in the abdominal vifcera; where, happily for him, the
fecretions were foon increafed fo as to vent off
the redundant humours? Let thofe who can
judge of the laws that obtain in the animal
œconomy determine this queftion. ———

Perfpiration can never be intirely ftopped
in a living body. The reafons affigned for
difeafes from an abatement of it are, that not
only a plethora is thereby brought on, but the
humour itfelf is fuppofed fo putrefcent as to
make it in the higheft degree dangerous to have
any part thereof detained, even for a fhort
time. Were it, indeed, true, that five-eighths
of

of our food and drink ought to pafs that way; and were it poffible that this difcharge could be fuppreffed for a number of days together, the appetite continuing good all the while, alarming overfulnefs undoubtedly would enfue, unlefs the other excretions increafed according as the perfpiration diminifhed. But we are well affured, that for much the greateft part of the year the common quantity of this difcharge falls far fhort of the above proportion in *Carolina*, where the climate is warmer than in *Italy*: *Sanctorius* himfelf found it fo for three months; and he might have extended the time ftill farther, as already hath been faid.

That there ought to be fomewhat of a regular proceeding in this difcharge as requifite to health, no one can doubt. And we may fay the fame of all the other excretions, which fhould have certain proportions, to free the body in different ways, of fuch things as are acrid when firft taken in; became fo afterward; or, having ferved the ends for which they were admitted, ought then to be expelled, to make way for frefh fupplies. As to the quality of that humour which paffes the pores of the fkin, thofe who confide in the doctrine of *Sanctorius*, would do well to diftinguifh the notion of putrefaction from putrefcence: for, fome part of the milk only excepted, all animal fluids are putrefcent under

certain

certain circumftances, though a tendency thereto can never happen, fo long as they circulate regularly; for this implies a good condition of the whole fyftem. But why this fubtileft of all fhould be the moft fo, and of courfe difpofed to infect the reft of the fluids immediately, more efpecially in a conftitution that, perhaps, but an hour before, was in the beft ftate, does not clearly appear. We know that the thin excrements may be again abforbed, without ill confequences, though they had been mixed with things that were nearly putrid in the inteftines. Nor can it be with reafon conceived, why the perfpiration fhould be more liable to putrefaction than the urine, which, though it confifts rather of groffer and perhaps more corruptible parts than the infenfible difcharge, produces no bad effects when abforbed and mixed with the blood. Of this any perfon may be prefently convinced, by exercifing fo brifkly, as to make himfelf fweat for an hour or two; when little urine will remain to be voided, though the bladder had been fo full before as to occafion a ftrong defire to empty itfelf.

When the urine is analized, it yields, 1*ft*, A tartareous falt, compofed of an alkaline earth, and a fmall quantity of oil and an acid, which is found adhering to chamber-pots, that are not duly cleanfed. 2*dly*, An ammoniacal falt, which is generated from a mixture of volatile
with

with acid falts; fo that, if a little alkaline falt, or quicklime, be added to infpiffated urine, a volatile urinous fmell will immediately enfue; as when thofe ingredients are mixed with crude fal ammoniac. 3*dly*, A fixed falt, which is partly alkaline, and in part neutral, as commo nfalt is. *Vide* opera *Hoffmanni* in folio, p. 69.

We fhall by-and-by enquire, whether the perfpiration and urine be not nearly, if not quite the fame, except with this difference, that the infenfible perfpiration may not contain fuch grofs, faline, fulphureous and earthy parts as pafs off by urine. Whence then does the latter derive fo deleterious a difpofition: for all thofe contents exifted before in the fluids, when the perfon was in health; and the like principles are alfo found in the other excretions, of the beft regulated conftitutions?. Neither will the fuppofal avail, that, as being a humour of the laft concoction, it, for that reafon, muft be the moft putrefcent: for, even in this refpect, it is only on a foot with the other natural difcharges; which being feverally fecreted and excreted in the due order of time, muft therefore be perfectly concocted. It ought, however, to be obferved, that thofe contents of the urine only appeared in the above manner after they were concentrated, and perhaps heightened by the great heat ufed in fuch proceffes; but were much lefs active,

when

when diffufed throughout the circulating
fluids.

The difcharge by perfpiration is, perhaps,
the only one which cannot be wholly ftopped
for feveral days together without very dangerous confequences. But the retained matter
here, as well as the urine, may be taken up
again by the veffels, and circulate a-new with
the fluids, as already hath been faid. In
dropfies, be the patient's drink ever fo plentiful, little urine will be excreted in proportion ; fo that whatever ought to have paffed
that way, muft in fuch cafes be lodged in the
cellular membrane, or cavities of the body,
together with the ferous parts of the blood,
where they may continue ftagnated, for fome
months, without fatal effects. When, therefore, a patient recovers from a dropfy, nothing
feems more plain, than that the humours
which had lain fo long in a ftate of reft, muft
firft be abforbed, before they can either be
expelled through any of the excreting organs,
or affimilated again in the courfe of circulation, as the veffels recover ftrength. It is
needlefs to fay much of thofe extravafations
that happen from contufions, in which the
blood that had for many days been deprived
of motion, is again diffolved and abforbed,
without any injury, though the contrary might
be expected ; for here, as well as in a dropfy,
fome degree of putrefaction might be looked-

for.

for. But as there are fmall and oftentimes
no evidences at all thereof on thofe and many
other occafions, I have thought that term was
frequently made too free with in fpeaking of
difeafes.

Surely no man in his fenfes, will believe
that the perfpirable matter can poffibly be fo
acrid or cauftic as that which gives the Small-
pox. But though inoculation informs us of
the inftant of time the fluids are infected, yet
it commonly requires fix or feven days, before
they are fo generally tainted thereby as to produce a fever, in order to expel the morbifick
caufe : And the fame, probably, may be faid
of other contagious difeafes, though we cannot always afcertain the exact time when the
patients received the firft feeds of thefe diforders. But if a perfon in health happens to
ficken prefently after going out of a warm
room into the open air, the mode is to impute
the diftemper to a check of the perfpiration ;
whereas in truth, we ought to look for another
caufe for fuch fudden violent effects.

But were the matter we perfpire really fo
difpofed to putrefaction, it might with reafon
be expected, that having brought on a fever
of many days, it would in that time become fo
putrid, (from the many circumftances which
then concurred to make it fo) that it would
give ftronger proofs thereof than it commonly
does.

does. Yet we are feldom fenfible of much a-
crimony in the fluids, as they pafs fo many
fine irritable canals in their way to the fkin,
either before or during the critical fweat;
though that is fuppofed to free the body of the
morbifick caufe, which, according to the
Sanctorian doctrine, confifts of fomewhat very
acrid, or deftructive of health. To me it feems,
that were the perfpirable matter really fo vi-
cious in its own nature, as immediately to
difpofe the fluids to putrefaction, its acrimony,
or whatever elfe fuch mifchievous properties
are derived from, would be fo heightened by
a fever, that the fweats, fo much wifhed for
at thofe times, could not happen, becaufe of
the conftrictions it might occafion on the fine
exhaling veffels by its irritation. Nor would
the urinary organs be exempt from the like effects. So that the expulfion of the febrifick
matter would be prevented either way.

The advantages arifing from poultices and
fomentations, and the fpreading of contagious
difeafes, among other things, plainly prove,
that we receive the fteams which exhale from
other bodies. It is alfo true that each individual perfon in a large company, may, without
any prefent or future inconveniency, imbibe
fome of the perfpirable matter which was difcharged from a great number of people, of
very different conftitutions, even after it had
been diffufed in the moft promifcuous man-

ner

ner through the heated air of a room. But when any perſon ſickens, after being in ſo crowded an aſſembly, the reaſon commonly given is this plain, and no leſs true, one, that he was injured by going out of a warm apartment into the cool damp air, heated as he then was.

But, there are fevers which attack when no ſenſible abatement of the perſpiration can be impeached : on the contrary, it proceeded freely, ſometimes too profuſely, before they came on, and continued plentiful, or with ſhort interruptions only, throughout thoſe diſeaſes. And we likewiſe know, that this diſcharge is ſmall, for ſeveral months together, at ſome ſeaſons, with great advantage to many. On the whole, we have good grounds to believe, that neither this, nor any of the other natural excretions, are limited to exact or ſtated rules, but vary not only in different people at the ſame time, but even in the ſame men daily, who yet enjoy health.

Dr. *Lining's* tables ſhow, that when the urine was greateſt perſpiration was always leaſt, and the contrary. Hence it ſeems, that in countries, where hot and cold weather ſucceed each other, the fluids are determined in a fuller ſtream toward the ſkin at one ſeaſon, and are again more repelled into the viſcera at another. For this reaſon, the diſcharges from within,

or

or through the outward pores, muſt increaſe
or diminiſh by turns, according as the humours
are more or leſs directed one way or the other.
The difference therefore between the urine and
perſpiration might not be great at the year's
end, as Dr. *Lining* diſcovered, though the
exceſs was in favour of the former, becauſe
the weather was uncommonly cool during
the twelve months he made his experiments.

As ſuch a conformity appears in the quantity
of thoſe two diſcharges, the augmentation of
one ſeems intended to make up for any defect
in the other, which muſt of courſe happen on
a change of the weather. They alſo are the
beſt ſuited to anſwer the great end of becoming ſubſtitutes for each other, as being the
moſt conſiderable of all the excretions. And
though one may not increaſe exactly in the
ſame proportion as the other leſſens, yet they
ſeem to act reciprocally as counterpoiſes, when
the weight of the fluids and ſtrength of the
veſſels are pretty well adjuſted.

To ſhew the truth of what hath been ſaid,
of the difference in thoſe two diſcharges during hot and cold weather, let us take the
medium of the perſpiration for twenty-four
hours in July, and compare this with its
mean quantity for the ſame ſpace of time in
February, they will then reſpectively be as
eighty-ſeven to thirty-ſeven. A plain proof
how

how greatly the perfpiring veffels were dilated, and how much more plentiful the flow of the fluids towards the outward arteries was in the Summer than in the Winter. Again, if the mean quantity of the urine for twenty-four hours in February be compared with its medium alfo for the fame time in July, they will be as feventy-eight to forty-four nearly. The difference therefore between the greateft excretion by perfpiration in the Summer, and by urine in the Winter, does not exceed nine ounces in twenty-four hours: But the difference between the fmalleft by perfpiration and leaft by urine in the moft oppofite feafons, was no more than feven ounces. An amazing correfpondence! though to the unthinking part of mankind, thefe things pafs unheeded, as if they bore no fort of proportion.

But a greater agreement ftill appears between thofe two fecretions than refults merely from fuch a conformity in their quantities, by which they feem intended to fupply each others defects at different times of the year. And if they really do fo, as is likely beyond a bare probability, it ought to follow, that the matter which paffes both ways, fhould be nearly, if not quite the fame; and that it is fo, an eafy experiment will fhow. For, if a fmall quantity of the perfpiration from the lungs, be collected by blowing in a clean wine glafs, and fome urine is made in another, (both be-

C   ing

ing taken fasting, or before the humours have
been diluted with fresh *chyle*), any person having an acute taste and smell will at once pro-
nounce that both are urine. Should a little
salt of tartar be added to each, a slightly volatile, but not disagreeable, smell will arise from
both : and if they be permitted to stand for a
day, without having any thing mingled with
them, a more acrid taste and scent will then
be perceived ; the urine, however, being still
most pungent (1). Moreover, nothing seems
more plain, than that those salts which pass
with the urine in temperate seasons, are discharged through the skin when the weather is
very warm : for black gloves which have been
worn, in such a season, till they were wet
with sweat, will be covered with a crust of
white salt when dry ; and our sweat also has
the same saltish taste. What we then perspire
smells at first as fresh blood ; but after some
hours, it communicates a urinous scent and
yellow colour to our linen : So that if those
things do not amount to a full proof, they

(1) Besides the general perspiration from the greatest
part of the outer surface and the lungs, particular drains
seem appointed to free the neighbouring parts of some
more highly saline and sulphureous matter. For what
perspires from behind the ears, from the gums, arm-pits,
groins and between the toes, hath a rancid offensive
smell, which differs in each place. So that could the
whole perspiration be collected together, there is no
room to doubt, it would make a mixture, full as acrimo-
nious as the urine.

make

make it extremely probable, that the perspiration and discharge by urine free the body
in different ways of much the same sort of
matter; and that what perspires from the outward surface and lungs is the same as urine
in the kidneys. The pores of the skin are so
many channels, intended to clear the external
parts of whatever hath become useless and un-
fit to be detained there. The lungs, and paf-
sages leading to them, perform the same of-
fice for the breast and neighbouring parts ;
the kidneys and intestines doing the like for
the lower viscera : so that, according to their
various situations, they severally contribute to
the same great ends. And, though the chief
excretions seem to go on without any esta-
blished rules, in various climates, and in the
inhabitants of the same country, at different
seasons, or even in the same persons from day
to day, (some making more urine, whilst lefs
passes by perspiration, and *vice versa*) yet all
of them may enjoy health if the requisites of
their several constitutions be but answered
either way.

Were it indeed true, that a plethora from
an abatement of the perspiration must of courfe
happen before inflammatory fevers can attack, it might be expected that fuch disorders would be most common in the Summer,
because a much greater quantity thereof may
be detained in a short time, that discharge be-

ing

ing then greateſt, and the fibres very lax and
irritable, which renders them ſuſceptible of
ſtrong ſpaſmodick contractions from ſlighter
cauſes. It, however, falls out quite other-
wiſe; for ſuch diſeaſes ſeldom appear with us
at that ſeaſon, but moſtly in the Winter and
Spring, when perſpiration is little, and, con-
ſequently, leſs likely to produce the overfulneſs ſo much contended for. But at no time
can ſuch an addition be made to the circulating fluids thereby, in any equal number of
hours, as by eating and drinking plentifully,
and even intemperately, according to the daily
cuſtom of many, without much inconveniency, at leaſt, for the preſent, though ſeveral
pounds of freſh chyle, the produce, perhaps,
of an ill-judged mixture of meats and drinks,
paſs into their veſſels oftener than once in the
day. It may probably be ſaid here, that the
excretions are ſtill going on, and, therefore,
whatever overfulneſs may happen from ſuch
exceſſes, are abating every moment. But is
not this equally true on a diminution of the
perſpiration? for though the outward pores
may be occaſionally ſtraitened, yet that diſcharge is ſtill proceeding from the lungs, and
perhaps more plentifully, becauſe a greater
quantity of blood muſt circulate through that
organ in all fevers, ſuppoſing no hæmorrhage
or any exceſſive increaſe of the abdominal ſe-
cretions happens in the mean time; beſides,
that urine alſo is commonly augmented at ſuch
times :

times: fo that perfpiration is the only difcharge whofe defects can be compenfated by
another channel. It hath been faid, that no
great plethora can happen from an intercep-
tion of the perfpiration; yet, for the fake of
argument, let us even grant, that fomewhat
of an overfulnefs may at times enfue from that
caufe. But then might we not expect this
would be removed, and that the fever it oc-
cafioned ought to ceafe, after repeated bleedings, fweatings, and the like evacuations fo
commonly made in fuch cafes; more efpecially as the appetite fails for every thing, except drink or thin nourifhment, which foon
paffes off. Neverthelefs the difeafe too often
continues, though the patient is fo emptied
that he can fcarcely be known. Befides, nothing is more notorious, than that the body
fometimes is exceedingly wafted by a fever of
a few days, though no artificial evacuations
had been made, and little paffed by any of
the fenfible difcharges; the patient alfo having all this while been well fupplied with
fuitable diet. No way then remains to account for this great lofs of fubftance, but by
fuppofing the perfpiration to have been vaftly
increafed, from the lungs moftly; for the fkin,
tongue, and throat, are at times dry and parched on thofe occafions.

Whether thofe things will be confidered in
the fame light by others, I know not: but to

me

me it feems, that whatever plethora may enfue, from a temporary and partial check of
the perfpiration (for partial it muft be), cannot of itfelf fo often bring on a fever, as already hath been faid, if it can at all. For, according to *Sanctorius* himfelf, were the whole
of it detained, it would only make an addition to the fluids of one pound in twenty-four
hours from the autumnal equinox to the Winter folftice ; or of thirty-feven ounces in the
fpace of one day and night in February, according to Dr. *Lining* ; fuppofing ftill, which
yet can fcarcely be, that the whole of this
were matter that ought to have perfpired,
without any addition being made to it by the
moifture abforbed from the air. However,
this is only mentioned in order to give the argument its full fcope, but without admitting
that an entire ftop can poffibly be put to the
perfpiration, fo long as the humours circulate
at all.

If a dangerous overfulnefs cannot well arife
from a detention of the perfpiration, in the
common courfe of things, as little is its fuppofed pronenefs to putrefaction likely to bring
on a violent difeafe fo fuddenly as oftentimes
happens to people who, but a few hours before, were healthy and ftrong ; for furely it is
difficult to conceive, how either the quantity
or quality of that fluid, which might occafionally be intercepted, coux excite fuch di-
ftreffing

ftreffing and hazardous effects in fo fmall a
time. I would afk thofe who maintain the
doctrine we have been cenfuring, Whether it
can fairly be imputed to an abatement of the
perfpiration, that a perfon on going out of a
warm room into the cool moift air, or that the
man who drinks cold water, or fits down on
the damp ground when he is much heated
with exercife, fhall immediately fhiver, and
have a fever, pleurify, or quinfy? that one
who plunges himfelf into the water whilft he
is fweating freely fhall inftantly become rigid,
infenfible, and die in a few hours?—or that
thofe who are expofed to rain in the Summer,
or fleep only for a little while out of doors at
night during that feafon, fhall prefently be
cramped all over, and have a tetanus, or fome
other dangerous difeafe? Many inftances in
all thofe ways have I known. And fuch effects
ought undoubtedly to be afcribed to fome
other caufe, than a diminution of the perfpiration alone. We alfo know for certain, that
this difcharge undergoes remarkable checks,
for many hours together, without much inconveniency, as almoft every man can recollect to have happened to himfelf on various
occafions

CHAP.

CHAP. II.

Of the immediate Cause of Fevers.

THOUGH it be not the lot of man to enjoy conftant health, many perfons however are happily endowed with fuch well-conftructed conftitutions, that, for a number of years together, the veffels every where have proper ftrength, and continue fo flexible that the fluids they contain, can pafs freely from the heart to all parts of the body, and return without any hinderance to that organ. Under circumftances like thefe every part receives only its natural proportion of the blood: the feveral fecretions and excretions will be made in regular order; and as no redundance can happen, fo nothing will fall fhort, till the equality of action and re-action that fubfifted between the folids and fluids is interrupted: and then the natural, vital and animal functions muft be difordered in proportion. It therefore feems plain that when any man is attacked with a fever, in whom the conftitution was fo well difpofed immediately before, the fault could not have been in the fluids, (for they neither exceeded the quantity, nor erred in quality) but in the veffels alone, owing to fome change in the condition of their coats:

For,

For, beyond all doubt, they are not now in the fame ftate as before; and according to the alteration they have undergone, fo muft more or lefs of diforder enfue in the diftribution of the blood, and confequently in all things that depend on its regular circulation.

I know but two ways in which the equality of mutual action between the folids and fluids can be broken in upon, or, in other words, that a difeafe can happen; namely, when the veffels are fo diftended by an unnatural rarefaction of the humours, that they cannot circulate freely : but fuch an expanfion feldom occafions a diftemper in temperate climates; and muft then be owing to too much heat, a clofe confined air, or fuch as does not gravitate fufficiently, whence fudden death fometimes enfues. Nor does any thing prevent the blood from being expanded by the natural heat of our bodies, fo as to deftroy us: this is checked and kept within bounds by the fuperiour fpring of the veffels.—Diforders of this fort, however, not being the fubject of our prefent enquiry, we fhall pafs to the other, which is by far the moft common caufe of difeafes.

It is very fingular, that almoft every fever fhould be preceded by a fenfe of coldnefs in fome degree; and the greater this, and the longer its continuance, the more dangerous

rous for the moſt part is the diſorder. This chillineſs may come on ſuddenly : but, more frequently, the patient hath ſucceſſive chills and fluſhings, perhaps, for ſome days ; and in the mean time, an irkſome ſort of weari-neſs is perceived, chiefly about the legs knees and thighs, together with a liſtleſſneſs of the other parts, and of the mind alſo. Yet, all this while, the pulſe, though ſomewhat hard, may not be much quickened. Thoſe ſenſa-tions of coldneſs, return oftener and continue longer, more eſpecially if the perſon expoſes himſelf to the wind ; and the other ſymptoms likewiſe increaſe, till at laſt a ſtrong ſhivering or horror comes on. At this time the hands and feet are cold; the veins which lie immediate-ly under the ſkin diſappear; and the face, to-gether with all the other outward parts are pale, and, as it were, ſhrunk or contracted. An inſatiable thirſt attends ; a painful ſtiffneſs is felt in the joints and all the external muſcles, but moſt ſeverely from the hips downwards. And as the fluids are now made to accumulate in the internal and ſuperiour parts, a ſickneſs at the ſtomach, anxiety, quick-breathing, and head-ach enſue, with great heat about the epigaſtrium, breaſt and forehead, which after ſome time diffuſes itſelf over the body and limbs, if proper care be taken. The pulſe is confuſed, ſmall, hard and very quick during the horror ; but it be-comes ſomewhat fuller, more ſoft and ſlow
when

when the cold fit ceafes. Though it is not indeed uncommon for the fick to be more or lefs chilly at times throughout every acute difeafe, at all feafons, more efpecially on turning in bed, or when the clothes are lifted up, fo as to admit the cool air to the fkin.

If I judge rightly, the above fymptoms clearly fhow, that the motion of the fluids is in fome meafure intercepted in moft, if not all, the outward veffels and mufcles: the fmall contracted pulfe, points out fome fault in the coats of the arteries, which are too much braced to permit the blood either to enter or pafs them freely; and the palenefs of the face, and fhrinking of all the outward parts, likewife make it evident, that the fkin together with the contiguous membranes are fpafmodically affected. It is no lefs worthy of obfervation, that this conftriction took place in fome degree before any perceivable quicknefs happened in the pulfe, and, likewife, that from the time it came on, even in the flighteft manner, health was impaired in proportion. But the bad effects of it appeared more clearly in the time of the fhivering; for the contraction having then not only become ftronger but alfo more general, by communicating itfelf to a great number of veffels, and other membranous parts, whatever order fubfifted before, in the natural, vital, and animal œconomy, was thereby confounded. If thofe
things

things be true, as they certainly are, may we not fafely conclude, that the immediate caufe of feverifh complaints, is an unnatural conftriction of the arteries in particular; and that the ftronger this, and the greater number of veffels it affects, the more fevere muft be the diftemper?

The outward branches of the defcending aorta are, probably, moft expofed to accidents of this fort from cold or moift air, as being diftributed through much the greateft part of the body, and conveying the blood to places moft diftant from the heart, where its velocity and momentum are fmall; for which reafon fuch conftrictions may the eafier be brought on them. And we accordingly find, that the heat fails firft and moft in the legs and feet; and the painful ftiffnefs is alfo more perceiveable there than in other parts.

Whatever can give much pain, or ftimulate the nerves, fo as to caufe them to excite fuch conftrictions, may bring on a fever. But we do not fuppofe all the arteries muft be thus affected, to produce the difeafe, it being fully fufficient for this purpofe, that fo many of them fhould be fo difordered, as will occafion an irregular diftribution of the blood; and that they be kept in this ftate to continue the diftemper. Nothing is more known, than that there are plain figns of great laxnefs in fome

veffels,

veſſels, whilſt, at the ſame time, others are
apparently too much braced; witneſs the ſmall
hard pulſe, and brown parched tongue in
choleras and fevers with vomitings, purgings,
profuſe ſweats, &c.

It will probably be admitted, that ſome
degree of ſpaſm, does, indeed, take place in
time of the horror, which muſt be removed,
before a general warmth can return, more eſpecially ere the patient can ſweat, or the urine
become thick. But we are well aſſured, that
diſtreſſing ſpaſms of particular parts oftentimes
happen, and yet only a hardneſs, without
much quickneſs, will be perceived in the
pulſe, as is found in nervous and other cho-
licks; the head-aches, choakings, and ſome
other complaints, common to hyſterical and
hypochondriacal people ; in the biliary duĉts
during a jaundice; and even in the Tetanus
itſelf, except during the general temporary
contraĉtions which are common to that cruel
diſeaſe. So, in endeavouring to eſtabliſh the
above theory, it was no where inſinuated, that
the arteries muſt be ſo conſtriĉted, as wholly
to prevent the fluids from paſſing; though
this ſeems nearly the caſe ſometimes in thoſe
of the extreme parts, during the ſevere horrors of quartan and tertian intermittents. But
that is a ſtate of ſuch violence, as no con-
ſtitutions can ſupport for many hours together:
and we accordingly find, that they who die of
thoſe

thofe difeafes, never recover a proper warmth
after the horrors attacked them; either a
mortal anxiety, lethargy, convulfions, or an
apoplexy foon enfuing. 'Till the fatal ftage
of a fever, or probably juft at the inftant of
death, this fpafm may be ftrongeft in time of
the cold fit; but notwithftanding the force of
it does abate (for abate it muft) before the
body can every where become warm, the
action of the heart being now weakened, and
its motions much hurried, by the præcipitous
return of the venous blood (1), it neither can
receive nor throw out fuch a quantity of that
fluid, nor communicate fuch an impetus to it,
as is neceffary to dilate the conftricted arteries
fully; and they, in their turn, not only re-
fufe a ready admittance, but a free paffage alfo to the blood, as already hath been faid.
Befides, the coats of the arteries being ren-
dered more elaftick from the time the fpafm

(1) Our moft gracious Sovereign King GEORGE II.
in all likelihood, could not have died fo fuddenly, from
the burfting of the right ventricle of the heart, (thin as
all the veffels were faid to be) unlefs a general forcible
fpafm of the coats of the arteries and other mufcular
membranes had preceded, though no perfon happened
to be prefent when the horror attacked. The caufe of
his Royal Highnefs the Duke of *Cumberland*'s death, was
more evident; for he underwent two or three fucceffive
ftrong rigors or fhiverings, whereby the fluids were fo
impetuoufly urged into the internal and fuperiour parts,
that the veffels of the brain were made to give-way to the
violent irruption of the blood, which was thus forced
upon them.

came

came on, and ftill more fo by the horror,
they will maintain the fuperiority they have
already acquired over the feeble action of the
heart, and the little refiftance the fluids can
make: the difeafe therefore will not only continue, but may even increafe, if the caufe be
not removed.

The fame reafoning will hold exactly in
thofe fpafmodical contractions that arife from
the irritation of fome fort of acrimony lying
in the ftomach or inteftines : and they alfo
will continue till that be either corrected or
expelled. As to the occafional fweats, and
thick urine in fevers, it hath been faid, that
the fpafm needs not be very general to bring
on and keep up thofe difeafes ; for a perfon
may be dangeroufly ill, though only part of
the fyftem is thus affected ; and even in the
opifthotonos itfelf, fome mufcles commonly
are capable of voluntary motions, whilft others
are detained in an unvaried ftate of violent
contraction. Nothing farther can be inferred
from thofe fweats in fevers, than that the outward veffels are not fo ftrongly conftricted, as
wholly to intercept the fluids from paffing to
the exhaling canals; or that the fpafm was at
fuch times fomewhat abated : and the thick
urine fhows the fame ftate of the fmall emulgent arteries. But, though thofe unnatural
contractions may abate in part, they cannot
be faid to have ceafed, fo long as the pulfe is
not

not foft, be it ftrong or weak, quick or flow ;
this hardnefs being a proof of too great fprin-
ginefs in the arteries.

Having offered fome reafons, which, at
leaft, feem to favour my opinion of the immediate caufe of fevers, it may not be amifs
to point out a few of its moft obvious effects:
and thefe, with others not here mentioned,
would be better underftood, if the notion of
fome anatomifts were true, that the feveral
different orders of veffels confift only of various arrangements of one and the fame continued canal, which is every where full of
fluids. Nay, farther, all the flexible parts of
our bodies, are alfo included in the above fuppofal; of which were we affured, it might
then be eafy to account for the fympathy that
evidently takes place between one part of our
bodies and another.

Firft, We obferved, that from the time the
fpafm began, and whilft it was gaining on
the veffels, the equality of reciprocal action,
which fubfifted before between them and the
fluids, was interrupted; and health alfo de-
clined in proportion to the force and extent of
that contraction.

Secondly, As the blood was tumultuoufly
pufhed forwards in the veins at the time of
the horror; becaufe of the ftrong preffure that
                                        was

was then made on them, by the conftriction
of the fkin and other mufcular membranes,
(the coats of thofe veffels being likewife ren-
dered more fpringy thereby,) it is plain, that
the heart ought either to have admitted and
expelled more blood in the fame time than it
does in health, or performed its motions
quicker. But the former not being poffible,
(it being even fcarcely probable that it could
receive fo much, as its mufcular fibres might
be fpafmodically affected in fome degree,) the
latter unavoidably happened.

Thirdly, Becaufe the capacities of all the
arteries, on which this fpafm acts (1.), muft
be leffened in proportion, fo they can neither
receive nor tranfmit their natural quantities
of the fluids, fo long as they continue in that
ftate : An overplus muft therefore be admitted by others beyond what they ought to
contain, were the circulation every where free.
And the ftronger this check in any confiderable number of veffels, the more muft the
blood be collected, and the greater its impetus
in thofe that are more open and paffable, as
being but little or not at all affected with the
fpafm; unlefs where fuch overfulneffes happen, as difable the arteries from contracting
themfelves properly.

Fourthly, From this obftruction (1. 2. 3.)
to the free and equal diftribution of the fluids,

D   fome

fome ftop muft be put to the fucceeding blood:
this again will be communicated to that which
follows ; and fo on, till fo much as cannot
pafs the conftricted veffels, is made to recoil,
by an inverted fort of circulation, on thofe
that have not undergone any unnatural contractions, or on thefe that are leaft able to
refift its weight and impulfe. The overplus
will, therefore, fall chiefly on fuch veffels as
are naturally weak, or on thofe that are not
fupported by furrounding mufcles : And we
accordingly find, that thofe of the vifcera and
brain, are always overcharged in fevers, if
nothing happens to vent the redundance,
which elfe muft enfue in them.

Fifthly, Under circumftances like thofe (4.)
the veffels that are thus too much diftended,
cannot clear themfelves of this additional
quantity of the blood: partly, becaufe they
are fo ftretched thereby, that they have not
the power to complete their fyftoles ; but
chiefly, becaufe the veffels in other parts are
not at prefent in condition to receive their full
proportions of the fluids.

Sixthly, When things have come to that
pafs (5.), fuppofing the fpafm ftill gaining, and
urging the fluids yet more into the internal
and fuperiour parts, the patient's ftrength being at the fame time greatly fpent, the blood
will then be compelled to retire within a nar-
row

row compafs: and, at laft, being moftly collected in the brain, lungs, and large veffels
near the heart, the fmall arteries in the outward and extreme parts fhut themfelves up ;
the arms and legs lofe their heat, and the mufcular flefh feels hard. The aortas no longer
receive the blood freely from the heart; this
organ can admit but little from the *vena cava*
and *pulmonaris*; and the lungs alfo, being already overfilled, receive only a fmall quantity
from the heart. But, here let us ftop! for to
thofe who are much weakened, the laft hour
cannot now be far off. Nay, when the like
fymptoms happen to them who enjoyed a
good fhare of ftrength but an hour before,
death muft foon enfue, if they cannot be
fpeedily relieved. They, therefore, who are
cut off by acute difeafes, may with truth be
faid to die a violent death : for the actions of
thofe organs on which life depends, are ftopped, and as it were overwhelmed by this exceffive accumulation of the fluids in them,
the veffels in other parts being very generally
clofed at fuch times ; fo that the patients yield
as victims to the vehemence of the fymptoms,
rather than becaufe their bodies are drained
of the refources of life. For in thofe diftempers that prove mortal in a few days, a fufficient quantity of fluids oftentimes remains to
the laft gafp to continue life, were the veffels
but in a condition to circulate them regularly.
On the other hand, when any perfon finks

under

under a hæmorrhage, cholera, purging, spit-
ting, or other sudden or flow discharges, the
body is, in the end, so deprived of its juices,
that a sufficiency of them is not left to fill the
remote small vessels ; which, for want of the
usual resistance to their contractiblenefs, and
because the projectile force of the heart is
then very small, shut themselves up to their
axes.

It is needless to recount more effects of this
suppoofed unnatural conftriction : for any others
that can happen in the courfe of feverish difeafes, may eafily be underftood from the fame
manner of reafoning.   And though, happily,
this fpafm does not always produce fatal confequences, yet, what we have related will appear in fome degree in every fever, unlefs the
fymptoms be abated by fome difcharge or
other, to prevent the overfulneffes that elfe
muft enfue, in the internal and fuperiour
veffels.

Should thofe things be true, may not many
fymptoms which are daily feen in fevers, be
eafily accounted for from the irregular diftribution of the fluids, that happens in confequence of this fpafmodick conftriction of the
arteries, whereby the blood is ftrongly re-
preffed from fome parts, and made to accumulate

mulate in others (1)? Before a fpafm attacks, the folids muft be over fpringy, and the blood too compact, otherwife a pleurify, inflammatory peripneumony, or a quinfy cannot come on.

It will probably be objected, that in thofe complaints the blood is evidently inflamed, as appears from the yellow or pale tough fkin, with which its furface is covered; the red part or grume being likewife more, and the ferum commonly lefs in quantity than in health. But fetting afide the confideration of the effect cold air may have in contributing to thofe appearances, when that fluid is expofed to it in a ftate of reft, we will even allow the blood to be really more denfe in inflammatory fevers; yet it is not lefs true, that it was by no means fo before its velocity had been un-

(1) *Hippocrates* feemed clearly to be of the fame opinion; for, in his chapter περι φυσων, he exprefly fays, " Efficiunt [horrores] ut in alia loca parcior, in alia
" jufto copiofior fanguis feratur; ex quo ftafes, ftagna-
" tiones, in partibus imbecillioribus productæ, graves
" et multas producunt ægritudines—Prohibetur fangui-
" nis curfus, atque alio quidem loco confiftit, alio len-
" tius penetrat, qua fanè inæqualitate tranfitus fanguinis
" per corpus facta, omnigenæ inæqualitates per omne
" corpus contingunt.—Sanguis [præfentem horrorem
" metuens] ab extremis corporis partibus, quæ exfan-
" gues, et propter frigus palpitant, decidit, et concurrit
" ad locos maxime calidos; et ob fanguinis ibi collecti
" abundantiam, fufcitari calorem, ad extrema demum
" exeuntem."

naturally

naturally augmented, and it fuftained a continued ftrong compreffion from the great
force the arteries acquired, together with
other confequences of thofe caufes. But, let
the conftrictions be removed, fo that a regular circulation may follow, the fever with its
painful fymptoms will then ceafe, and, in a
little time, no fuch vifcidity will be feen, becaufe the blood foon recovers its proper confiftence, after regaining a free and equal diftribution throughout the fyftem, and is no
longer under the like reftraints in paffing the
veffels.

It is commonly thought that perfons of
ftrong full habits are moft liable to inflammatory diftempers; but the effects of this
fpafm will be the fame in weakly people,
though no plethora took place before, provided the weather favours thofe diforders :
and the like tough fkin will alfo be feen on
their blood, but not in fo great a degree. I
have oftentimes known women feized with
fevere pleurifies, peripneumonies, or quinfies,
after being much weakened by an abortion,
or a flooding. The fame complaints have
likewife been brought on tender fubjects by a
colick: for the conftrictions, that in this cafe
arofe at firft in the ftomach, from the irritation of an acid acrimony, being communicated by fympathy to the outward fmall veffels, a horror thence enfued ; and whenever
that

that happens, be the caufe what it will, more or lefs of fever muft enfue ; the difeafe, moft commonly being of that fort which then prevails, whatever be the age or conftitution of the patient. Accordingly, we fee the ftrongeft men have intermittents, or fevers of the depreffed kind in July and Auguft : and on the other hand, puny perfons and infants alfo, are liable to inflammatory diforders in the Winter.

I will further obferve, that acute diftempers feldom continue with the fame violence and intenfenefs of heat: for very commonly the patients have either fenfible or obfcure chillinefs at times ; and then it may be juftly concluded, that the fpafm is ftill increafing, and the difeafe alfo in confequence thereof. At the time of thofe chillineffes, the pulfe is more quick, hard and fmall, and all the former fymptoms are heightened ; for fuch fhiverings truly are horrors, which not only aggravate the fpafm that acted before, but make it extend to a greater number of veffels, whereby the diforder will at leaft be prolonged. Such occafional fhiverings fhould therefore be prevented if poffible, by keeping the patients properly covered, according to the weather ; for every time the cold air hath accefs to their bodies, thofe horrors will more certainly come on, efpecially in the Winter.

If

If I have not explained myſelf clearly on the nature of the ſpaſm I meant to eſtabliſh, let us ſuppoſe it were poſſible to have all the veſſels of the human body laid open to view, the heart at the ſame time propelling the blood throughout the ſyſtem, and receiving it back again from the veins, as is regularly done in health. Here, were a preſſure made on any conſiderable number of arteries, ſo as to ſtraiten them in an equal degree with the morbifick conſtriction we ſuppoſe, the veſſels in other places would be ſeen to ſwell; and the more numerous thoſe that are preſſed, and greater the weight applied, the more would the fluids be made to retire thence and accumulate in others, where overfulneſſes muſt enſue, in proportion to the check given to the progreſs of the blood through the confined veſſels. All this is plain enough; and ſomewhat of the ſame ſort as certainly happens whenever the coats of the arteries are ſpaſmodically affected.

CHAP,

CHAP. III.

Of the Cure of common, continued, and inflam-
matory Fevers.

I HAVE for a long time thought, that,
among the variety of diftempers to which
we are liable, fevers, in particular, have been
divided into too many claffes : nor are the
ways of treating them lefs diverfified. It
might have been meant by this fhew of exactnefs, to leave us as little room as poffible
to miftake one difeafe for another; but, in
my opinion, that end had been better anfwered by fewer principles, well defined. For
experience hath convinced me, that it matters
not much, under what forms feveral acute
diforders appear, or whatever be the ages or
conftitutions of the patients (when no particular acrimony prevails), provided the complaints agree in fome circumftances with
others that are commonly believed to be of
very different natures; for no reafon that I
can perceive, but becaufe fome fymptoms fall
more on one part than another. Hence the
difeafe hath fome name which fuits well .
enough in converfation : but it fhould not be
thence inferred, that any material difference
ought always to be made in the manner of

curing

curing it, more efpecially at its firft attack.
For if it be granted, that a fpafmodic contraction of the arteries is the immediate caufe
of fevers, and that the fymptoms which happen in the courfes of thofe diftempers, are
owing as it were to an inverted or irregular
circulation of the blood, what hath juft now
been faid, will not appear chimerical; more
efpecially when it is applied to fuch feverifh
complaints as attack thofe, who but a few
hours before were in health.

It would be needlefs to give many inftances of a method I have for many years ufed
with fuccefs to remove common, continued,
or inflammatory fevers, within the firft or fecond day, when they were not attended with
a purging, which happens but feldom in the
latter fort. But, for an example, let us fuppofe a ftrong man to be attacked with a pleurify ; though this be as dangerous and di-
ftreffing a malady, as almoft any we are liable
to, it. will be removed in a few hours by
purging and fweating, if the difcharges be but
plentiful, and the patient be properly taken
care of. Nor will the reafon, why this management fhould have fo good effects, be a fecret
to thofe, who recollect what hath been faid
on the conftrictions of the external veffels,
and the overfulneffes they occafion within :
which, being only effects of the former, are
more directly abated by purging, than any
other

other evacuation that can be made; and, therefore, whatever inflammation or obftruction
might have enfued from the prefent diftended
condition of the internal veffels, will be prevented by thofe means. For great repeated
revulfions being thus made, as it were immediately from thofe parts, they will be relieved,
in proportion, of the overcharges they fuftained, by every loofe ftool. And, as the like
effects extend to all thofe veffels in which
any degree of plethora took place, and even
to the heart and lungs, each muft then act
with more power, fo as not only to clear
themfelves of whatever plenitude might ftill
remain in them ; but alfo, by communicating
a brifker impulfe to the blood itfelf, the fmall
contracted arteries will thereby be dilated by
degrees. Farther, to infure the patient's
fweating, hot bricks ought to be laid at a
convenient diftance from the feet and legs, to
affift in taking off the unnatural contractions,
that we fuppofed were ftrongeft thereabouts,
by the kindly warmth they give thofe parts.
The effect of this treatment is fuch, that, after the perfon hath had fome large ftools, and
fweated plentifully at the fame time, the pulfe,
which began to foften and fill, on the fecretions being freely promoted in the abdominal
vifcera, will foon become flow and natural, (a
proof that the fpafm is removed) and the blood
be circulated regularly ; which is all that was
required for the cure.

This

2

This is my common way of reasoning on the good effects of purging and sweating in the early stages of fevers: But whether I argue rightly or not, the advantages of such treatment have been confirmed to me by thousands of instances. To be diligent in observing what the same disease constantly requires to remove it, in the most speedy and effectual manner, and perfect our judgments therein, with all the certainty that experience and the nature of things will allow, is, in my opinion, the only true way to establish a rational theory and successful practice; as both would then depend on such fixed principles as must abide the test. Whether I have succeeded either way is left to others to determine : but this I may be allowed to say, that the above theory seems to justify the practice, as this does the former. For, beyond all doubt, had not the vessels within been too much filled to have increased the excretions directly from them, and have promoted so many plentiful stools, it must have had consequences of the most dangerous nature : and, on the other hand, had not the outward parts been too much braced or constricted, then, surely, to have brought on profuse sweats by relaxing them still more had been equally pernicious.

When I first began to use this method some disappointments happened, from my
not

not attending to that material circumſtance of
raiſing and keeping up a proper heat in the
legs and feet, and guarding the other parts of
the body from cold air, more eſpecially during
the winter: the neceſſity for which did not
then occur, though it was altogether conſiſtent
with the notion of a ſpaſmodic conſtriction.
But after care was taken that way few patients miſſed of relief, unleſs they themſelves
or their attendants thought it unneceſſary
to comply with ſuch ſeemingly trifling in-
junctions.

That the patients may not be obliged to get
up a warm bed-pan muſt always be carefully
conveyed to them under the bed-cloaths ; and
their drink and nouriſhment ſhould be given
either with a child's feeding-boat, or through
the ſpout of a tea-pot, as they lie covered :
beſides, when the bricks begin to feel cold
they ſhould be removed, and hot ones put in
their places, ſo long as may be neceſſary.
Though the fever, together with the painful
ſymptoms, will undoubtedly be abated by
thoſe means, yet they may not intirely ceaſe
on this firſt trial.   In that caſe the medicine
muſt be repeated, and the diſcharges promoted
more briſkly, unleſs the patient be already
very weak: but he muſt be ſo in an extreme
degree to deter us from attempting his relief
at once by purging and ſweating, rather than
ſuffer an expectoration to come on in a pleurify

rify or peripneumony: for at beſt, that gives only a chance to recover.

I muſt obſerve, that it is not always neceſ-ſary to repeat the laxative, though the ſymp-toms be not wholly removed, provided the moſt acute are abated. It may be ſufficient to mix ſome eſſence of antimony and ſugar, with a decoction of poppy-heads and anni-feeds in water, and to give the patient a com-mon ſpoonfull of it every half-hour or ſel-domer, (according to the caſe) till he ſweat freely, and the fever and pain ceaſe.—A dry tickling cough is ſometimes troubleſome for a few days after the diſeaſe is pretty well over. To allay this, let ſome of the inſpiſſated juice of liquorice be diſſolved in the decoction of poppies, then a little of the beſt olive-oil, in-corporated with the mucilage of gum arabic, be properly mixed with it, and a ſpoonfull at a time be given, as may be neceſſary.

It is evident, that the earlier this method is purſued, ſo much the better will it ſucceed. For if the breathing has been much obſtruct-ed for ſome days, by a ſevere pain in the ſide or breaſt, this, together with the ſpaſm which then acts on the outward veſſels, will cauſe the fluids to accumulate in the lungs; hence a dry cough muſt enſue from anxiety, at the beginning of thoſe complaints: but the ſecre-tions in that part being afterwards increaſed
by

by degrees, the phlegm that then lies extra-
vafated in the bronchial veficles will alf) excite a cough, by its irritation to expel itfelf;
and thus both the cough and fpitting muft
continue till the overfulneffes that took place
in the pulmonary veffels be removed, and the
profufe fecretions there ceafe in confequence
thereof. But thofe things would not have happened, had the complaint been abated within the firft or fecond day; at leaft they could
not be either dangerous or troublefome: and,
in general, the fame treatment anfwers very
well at any time before the expectoration becomes free.

The patient's chamber ought not to be too
clofe or warm ; nor his bed ftand very near a
fire, that the air he breathes may not be hot
and dry. His head and breaft muft be raifed
high ; and though he fhould be carefully covered, the bed-clothes need not be more than
he ufed in health, except from the waift
downwards, as the fweat procured by our method will be fufficiently promoted by the
hot bricks and warm drink and nourifhment,
which fhould be given often, but in fmall
quantities. For thofe purpofes, weak fage tea
fweetened a little with honey, weak broth
made with lean meat of any fort, and thin
water-gruel will ferve in lieu of every thing
elfe.

This

This hint for purging I borrowed from nature : for obferving when inflammatory difeafes were moft common, that thofe patients who had loofeneffes at the beginning of fevers, feldom were liable to complaints of the fides, breaft or parts above; and alfo, that the diftempers were generally mild and fhort, even in the winter; or, if a pleurify, peripneumony or quinfy came on during the prefent illnefs, this did not happen till the purging ceafed; and likewife, that if the belly became freely loofe, on the firft days of pleurifies or peripneumonies, they commonly were abated thereby, it feemed reafonable to me that a diarrhœa, promoted by art, would have good effects in fuch cafes; which accordingly fell out, and with greater advantages than were expected. But, to confefs a truth, at my firft fetting out in this way, I underwent fome ferious reflections on ftraying fo much from the common forms of practifing. Repeated purgings are recommended in inflammatory quinfies; but why fo carefully forbidden in the firft days of pleurifies and peripneumonies I fee no reafon. If this caution proceeded from fear either to delay or prevent the expectoration that of courfe is expected in thofe diforders, we, with equal probability, might apprehend the fame from the large repeated bleedings that are generally ufed.

What

What *Boerhaave* fays in his 830th and 888th Aphorifm is remarkably to the point: for, fpeaking of a peripneumony and pleurify, he tells us, ' *that a great difcharge of bile by ftool*' is one of the ways in which a crifis hap- pens in thofe diforders ; and his referve, ' *if the patient be relieved thereby*,' is as applicable to the other concoctions and excretions, by which he fays thofe diftempers are removed before the 4th day. How a fuitable increafe of the abdominal fecretions may relieve the veffels in the *thorax* of their prefent overful- neffes, and thereby prevent an obftruction, which might end in an inflammation, I eafily perceive ; but cannot fo readily comprehend the meaning of the word *Concoction*, fo com- monly ufed in fpeaking of fevers, without any eruption or tendency to a fuppuration (1).

Farther, if purging be difapproved at the be-
E
ginning

_____

(1) By *concoction* I fuppofe authors mean fuch a change wrought on the morbific caufe, as renders it either in- active and inert, or prepares it for being expelled the body by fome outlet. But in fimple, continued, in- flammatory, or other common fevers, to fpeak of mor- bifick matter and concoction, is more likely to lead us into a labyrinth, than fatisfy our judgments ; more ef- pecially at the beginning of thofe difeafes, when they attack perfons who juft before were in health. Indeed it is quite reafonable to apply thofe terms to contagious diforders, and alfo to fevers that arife from inflammations tending to fuppuration ; for there a morbific matter is evident. But I have known the greateft part of a large company ficken in the Spring or Autumn, after fpending
feveral

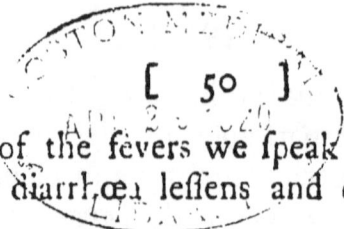

ginning of the fevers we fpeak of, becaufe a
plentiful diarrhœa leffens and even ftops the

feveral days together, on a party of pleafure as it was
called. Yet though they all were alike expofed to the
fame weather by day as well as in the night, and partook
of the fame intemperate exercifes and irregularities, yet
one perfon was prefently after confined by the gout or
rheumatifm; another had the quinfy; a third a dyfen-
tery, ague, purging, tooth-ach, and fo on. Here then
it is plain, that the fame general remote caufes acted
equally on them all; but it is no lefs true, they had as
various effects on their different conftitutions. Yet, according to the common mode of fpeaking, each complaint here had a morbific caufe proper to itfelf; the
nature of which let thofe explain, who, on all occafions,
adopt fuch reafoning. We fay with propriety, that the
eruption, when complete and benign, is a crifis to the
firft fever in the meafles and fmall-pox; whereas in fome
other acute diforders, there are certain exanthematous appearances, which never fail to portend death to the patients. Can any thinking man affert, that in pleurifies,
peripneumonies, catarrhal fevers, or pituitous afthmas,
the body is really difcharged of the morbific matter by
fpitting, which only empties the veffels that were overcharged? or when a fever, quinfy, tooth-ach, head-
ach, or even the diftempers juft now mentioned, are removed by purging and fweating properly ufed at their be-
ginnings, that this happened by our thus expelling their
morbific caufes. Or when the conftitution is much
relaxed, both as to the folids and fluids, by an intermitting fever, can it be faid, that the bark (which cures
merely by its bracing qualities, without promoting any
one fenfible excretion) then acts by expelling a morbific
matter of any fort? I would be underftood, that no di-
ftemperature exifted in the fluids, preceding the attacks
of fimple acute difeafes, on thofe who juft before were in
health; though it cannot be denied that fevers may bring
on fomewhat of this fort: when it is to be efteemed an
effect rather than the caufe of the diforder, notwithftanding it may require our attention to remove it.

fpittings

ſpittings in the ſecond ſtages of them, the effect complained of is a direct proof of the advantages of briſk purging at the time we recommend it. For nothing ſhews more clearly,
how ſudden and great a revulſion is thereby
made from the parts affected, by the humours being ſo haſtily carried off another
way : whereas, when thoſe complaints are
thus far advanced, ſuch a waſte of the fluids
ſhould by no means be made ; for they ought
rather to be detained then, to aſſiſt in diſ-
cuſſing or diſſolving whatever obſtruction or
inflammation hath happened in the lungs or
elſewhere. The objections, therefore, to purging on the firſt days of thoſe complaints, are
ſo far from having any weight, that, on the
contrary, they really ſerve to eſtabliſh the uſe-
fulneſs of it, when the veſſels within are only
overcharged or obſtructed in part, and before
an inflammation hath come on.

I follow the ſame method at the beginning
of every common fever when the belly is not
already looſe enough : for at ſuch times it is
by no means ſufficient that the patients have
ſtools as in health (which yet happens but ſeldom in the firſt days of inflammatory diſeaſes); becauſe, however well this might ſuit
when all the natural diſcharges proceed in due
order and proportion, ſomething more will be
neceſſary when this harmony is interrupted.
Nor will clyſters fulfill the intention, as much
experince

experience hath convinced me; not to mention, that, according to the foregoing theory, there are good reasons against the use of them.

The advantages of removing all fevers speedily, and in particular those of the sides and lungs, are too many and great to need many words. And it is needless to show how preferable the above method is to the common way of bleeding the patients, not only largely at once, but also repeatedly, which yet often fails to remove those complaints, probably, because some rules above laid down, are not observed, I mean, as to promoting sweats after such evacuations; for, let the patients lose ever so much blood, enough of that fluid will still remain to renew the pain if the spasm of the arteries be permitted to continue. Nay, it may even become stronger, and communicate itself to a greater number of vessels, the more the fluids are diminished, unless care be taken to prevent this; for the latter being the only antagonists to the strong disposition the arteries have to contraction, when this is not properly resisted, they will close themselves in proportion. Granting, however, that the disease be cured by a great loss of blood, this happens precisely by emptying the vessels, though in a less direct manner than we propose, considering the nature and seat of the complaint; but with this difference,

rence, that it weakens the patient much more
than purging and fweating. Suppofing after
all, that the diftemper is not removed by
bleeding, weakened as the patient already is,
his recovery will then depend on expectoration alone, which muft not only be plentiful,
but of a good quality alfo. A dangerous condition furely! for however well fuited this
difcharge may be, we have too many fatal inftances to prove, that it may be eafily fuppreffed by various accidents; befides, many ill
confequences often enfue from this manner of
terminating thofe difeafes. For thofe reafons
an expectoration fhould not be fuffered to
come on at all, if by any means they can be
cured without it. But it is worthy of remark,
that when recovery is effected even that way,
it is really compleated, at laft, in the manner
we propofe at firft, by emptying the internal
veffels at once; which is gradually done by
the fpitting, but with much more hazard to
the patient. On the whole, be the feverity
of thofe diftempers abated by whatever means,
a fweat will neverthelefs come on fooner or
later; when, and not before, the pulfe becomes flow and natural, a certain proof that
the fpafmodic conftrictions of the arteries are
then removed, as already hath been faid.

In fome countries bleeding may perhaps be
more neceffary in inflammatory fevers than
I have generally found it in *South-Carolina*;

yet,

yet, be this operation fo needful or not here, it is no where more practifed. They who are of opinion, that they ought not to depart from a rule, which hath fo long been as it were univerfally eftablifhed, may order fome blood to be taken away if they pleafe, provided the patient be ftrong; but I feldom do, unlefs the diforder has been of fome days continuance, or the pain in the fide fo acute as to obftruct breathing greatly, and thus occafion the fluids to accumulate in a high degree in the lungs; becaufe, for the moft part, much bleeding weakens the fick rather than gives lafting re- lief. For though the pain abates fometimes during the flow of blood from the vein, and may be eafier for a fmall time after, it will furely return, if the fpafm continues ; when every thing remains ftill to be done as much as if no blood had been loft. I therefore de- pend on purging and fweating, being morally certain of fuccefs that way, provided the dif- charges be plentiful, and things are properly conducted in other refpects.

By this method women have been cured of pleurifies and peripneumonies in every ftage of pregnancy, though inflammatory difeafes are known to be full of danger to fuch fubjects. One, in particular, being feized with an acute pleurify in the eighth month after conception, was effectually relieved in lefs than twelve hours : and fhe was as fpeedily recovered after

ter a relapfe, brought on by her going into a
wet room on the the third day after the firft
illnefs; though the latter attack was equally
fevere with the former: And feven weeks
after fhe was delivered of a healthy child.
Another, of a weak conftituticn, who re-
lapfed, twice into the fame diftemper in the
fpace of three weeks, was at each time fo foon
freed of it, that her ftrength was but little
impaired after having been thrice feized with
the pleurify within fo fhort a time. But, in
truth, the patients feldom are more weakened
by that diftemper, when treated in the above
manner at the beginning of it, than commonly happens from any other fmart fever of equal
continuance. May I be allowed to obferve,
that had thofe two women been managed in
the moft approved ufual way, the chance to
recover would have been much againft them.
I do not confine this practice to our natives
alone, but apply it to the ftrongeft Europeans,
when attacked with thofe difeafes on their firft
landing here in the winter: and the fuccefs is
alike in both.

I am not over anxious about the choice of
medicines; but, avoiding fuch as are very ir-
ritating, (of which there ftill is a preference)
I chiefly ufe manna, *Epfom* falt, elixir falutis,
Kermes-mineral rightly prepared, effence of
antimony, falt of tartar, magnefia, &c. which,
being varioufly combined, are given either in

mixtures

mixtures with water, or in powders, together with some drops of chemical oil, either of mint, or some other aromatic plant. But in whatever form they be administered, the ingredients ought to be so proportioned, that, by giving a small dose of the composition every hour, the expected operation may be brought on by the time the patient hath taken it four or five times: and whenever he begins to purge the medicine ought to be discontinued; for more stools will follow, which, together with the sweat that soon comes on, often remove the disease at once, or leave us little more to do. Every thing however ought to be given warm for some days; and the patient should lie in bed at least forty-eight hours, having hot bricks near his feet, more especially during the winter, to prevent his catching cold, which would surely bring on a relapse; for the solids being still very irritable, a spasmodic constriction will now be more easily brought on than if they had not so lately been affected that way. And, for the like reason, he must afterwards expose himself by degrees to the open air, in the middle of the day for some time, and then only, when the weather is fair and warm: but if it be cold he ought to be confined some days longer to bed, with hot bricks to his feet in case he perceives himself chilly, or those parts have not a proper degree of warmth.

Though

Though a tedious or dangerous fpitting will be prevented by thofe means, it muft yet be confeffed that a fmall cough and pain in the fide fometimes remain ; but without any other marks of a fever than fome degree of hardnefs in the pulfe : and this will more certainly happen, if the difeafe continued fome days before relief was fought. Thofe fymptoms, however, very commonly yield to fuch a decoction of poppies as already hath been mentioned : but the perfon ought ftill to be carefully guarded againft cold, and have all things given him warm. The fide, likewife, fhould be cautioufly bathed with a mixture of brandy, fpirit of fal ammoniac and laudanum (or opium had better be diffolved in the brandy) ; the part being afterwards covered with a flannel fqueezed hot out of the fame fpirit, and a warm tile laid over all. Or a plaifter of galbanum, mixed with a little opium and camphor, may be applied to the place.

I would remark here, how improbable the common opinion is, that too braced a ftate of the folids, and great compactnefs in the blood, fhould be fuch neceffary circumftances to difpofe us to imflammatory difeafes. For fuppofing a pleurify removed within the firft twenty-four hours from its attacking a ftrong man, nothing more frequently happens in this country than to fee it followed, in a day or two, by a quotidian or tertian intermittent ;

I   though

though difeafes of this fort, do not require
fuch a conflitution to favour their appearing,
as thofe of a contrary character are thought to
do. Nor do I fcruple to give the bark to
fuch patients, without regarding their having
been fo lately ill of an inflammatory diflem-
per; but then I begin with the decoction, and
add laxatives and diaphoretics to it, for the
firft two or three days.

Every phyfician of the fmalleft attention,
muft have obferved a remarkable difference
in the forms of difeafes, owing to the temporary changes our conftitutions undergo from
the weather, in the feveral feafons of the year;
and accordingly, the fpafm we fpeak of muft
have various effects. Thus we fee, that fome
of the excretions are at times much abated
thereby, whilft others are not increafed in
proportion: and, on the other hand, it pro-
motes exceffive difcharges, either by ftool,
urine, uterine or other hæmorrhages, according as the veffels in different places are difpofed to vent the fluids with which they
happen to be overcharged. And when the
refiftance to their flowing towards any one
part hath once been leffened by an increafed
difcharge made from it, they will rufh that
way in a freer ftream fo long as any degree of
fpafm continues on the arteries elfewhere. On
fomewhat of this nature depends the difficulty
we find in checking profufe fweats, diarrhœas,
catarrhal

catarrhal fpittings, and other irregular excretions of long ftanding, though no remarkable fever attends: becaufe the veffels whence they proceed have become fo dilated and weak, that their powers of contraction are loft; and therefore they cannot detain their contents.

When the fecretions within become unnaturally profufe in fevers, the fame regard is due to the fpafmodic condition of the outward veffels, as in the difeafes we have more fully fpoken of, by keeping the body duly covered, and laying hot bricks near the knees legs and feet, whilft the proper medicines are given to check thofe precipitate evacuations. But here, as well as in all other diforders, fpecial regard fhould be had to whatever acrimony may have taken place in the firft paffages, or paffed thence to the blood, whether it were an acid alone, or a mixture thereof with the bile, which, perhaps, might have brought on the difeafe, or continued it afterwards. In general, the diaphoretic regimen ought to be ufed here; and of the cordial bracing kind too, when the patient is much weakened and emptied by a catarrhal fpitting, cholera, bilious or ferous purgings or vomitings. But as too bound a belly is oftentimes the caufe of the latter diforder, fo it will not ceafe till the guts are freed of all hard excrements; for whilft one pellet remains within,

their

their periftaltic motions will, as it were, be inverted from the part it lodges in, and of courfe the vomitings muft continue. But maladies of this fort not being the fubject of our prefent enquiry, we fhall difmifs all farther confiderations of them.

Should any one ftill doubt that fuch fpafmodic conftrictions are, in truth, the caufe of fevers, let him but obferve what happens to the moft weakened perfons at the beginning of the fatal ftages of difeafes ; and it will then clearly appear, how the fame effects that arofe (though in a fmaller degree, becaufe the patients were ftrong) at the time this fpafm firft attacked, do indeed introduce the fcene that concludes the tragedy in all perfons who die of any diftemper. The certain approach of this mortal period is announced in many of thofe who are carried off by acute diforders, by more or lefs of a horror or rigor. But, notwithftanding the fame happens fometimes to them who have been worn out flowly by confumptions, purgings, or fome other difcharges of long continuance; and in thofe alfo who are fuddenly exhaufted by choleras, hæmorrhages, or otherwife, it for the moft part manifefts itfelf with equal affurance, by a gradual abatement of heat in the hands and feet. And, as the veffels there fhut themfelves up ftill more, this gains on the limbs in proportion, fo that they become cold and

fometimes

fometimes livid, the circulation being then at
an end in thofe parts. Thus the progrefs of
death on the body may be traced; the extreme parts being infenfible and inanimated
for many hours, perhaps, before life is extin-
guifhed in the brain and near the heart. On
the other hand, the death of thofe is more
fudden in whom a ftrong horror or rigor happened, becaufe of the general and violent
contractions of the veffels, which repel the
blood tumultuoufly, and as it were at once,
into the internal and fuperiour parts: When,
as the veffels within have not ftrength to refift
fo impetuous a torrent of the fluids as is then
made on them, and the fpafm at the fame
time gains faft on the arteries and other mufcular membranes, the brain, lungs, and heart
are fo quickly deluged and overwhelmed that
an end is foon put to the circulation of the
blood. In the courfe of my practice I have
fometimes been prefent when this rigor attacked, and always obferved, that, according
to the degree of it, death enfued fooner or
later. Though fome became immediately
infenfible, and could not move any part from
the moment the horror came on; others retained both their reafon and fpeech for a fhort
while, and complained either of a painful
ftiffnefs as if they were cramped, or a numb-
nefs of all the flefhy parts and joints, more
efpecially of the arms thighs and legs info-
much that they could not move them afterwards.

wards. On the contrary, fome cannot lie ftill at all, but tofs themfelves about in the moft reftlefs manner, fo long as they are able, from the dreadful ftifling which is caufed by the fluids being exceffively accumulated in the lungs and large veffels near the heart: Breathing is then very fhort, redoubled and fufpirious ; and the patient perceives an intolerable heat about the breaft, which feems to diffufe itfelf all over his body, fo that he will not endure any covering, and calls out to be fanned, and to have every window fet open, be the weather ever fo rigorous, though at the fame time he is bathed with clammy fweats and his legs and arms are cold as marble. The anxious reftleffnefs we juft now mentioned will continue fo long as the patient can move, or, as feems probable, till the veffels of the brain become fo diftended and full that they comprefs the origin of the nerves, when an end is at once put to all further fenfations. The brain being thus effectually furcharged, and the blood ftill continuing to be propelled that way, the like overfulneffes begin in the veffels of the cerebellum : and, from that inftant of time, breathing, which before was exceeding fhort, becomes very flow ; and the perfon moves no more, but dies in the pofture he lay when that happened. Infpiration is every moment performed at longer intervals now, and ceafes, as the laft act of life, when thofe veffels are ftretched and filled be-

yond

yond the power of exerting any farther contractions. Such rigors are indeed more common at the end of acute than chronical difeafes; but I have fometimes feen people recover of fevers, more efpecially from choleras, though the reftleffnefs and anxiety we
fpoke of feemed to indicate that death was
near. Yet, whether a diftinguifhable horror
came on or not, no one ever furvived this flow
breathing, under the circumftances above de-
fcribed; it being a proof, that the organ
whence the vital actions derive all their ener-
gy is overpowered beyond a poffibility of relief. For before that fort of breathing happened fuch an overfulnefs muft have been in
the veffels of the brain that they could receive
no more; the whole of the fluids being at
thofe times confined to few veffels.

In proportion to the progrefs of this fatal
fpafm on the arteries and other mufcular
membranes, and the retreat of the blood it
occafions, from the outward into the internal
and fuperior parts, How fpeedily do the
figns of death advance! With what anxiety
is breathing performed! How fluttering and
fmall the pulfe! and when it fcarcely can,
or, perhaps, cannot at all, be felt in the wrift,
what a ftrong throbbing may be perceived in
the carotids and about the epigaftrium! How
greatly are the veffels in the eyes diftended
(though they do not appear in health), which

gives

gives us reafon to believe thofe of the brain
are likewife in the fame condition! which
confequently is ever a bad fign in acute difeafes.——Whence proceeds that pale ghaftly
countenance, called a Hippocratic face ; and
why does it portend death to thofe who are
much weakened ? And how comes it that the
outward flefh fhould be hard and rigid at fuch
times, if no unnatural bracednefs took place
in the fibres even of the mufcles themfelves ?
General convulfions deftroy fome ; others
have only convulfive twitchings all over; fome
again have been watchful or outragioufly de-
lirious ; whilft thofe whofe bodies have been
drained of the vital fluids by degrees, are oftentimes fenfible almoft to the laft gafp. However, a peripneumony clofes the fcene in all
who die of any difeafe.

I fhall end this effay with fome queries,
which, in my opinion, cannot be fo well explained in any other way as by fuppofing the
fpafmodic conftrictions we mentioned to be
the immediate caufe of fevers, owing, probably, to fome fort of irritation affecting the
nerves in their courfes towards, or after they
are inferted in, the blood veffels and other
mufcular membranes. And it is not unlike-
ly, when a perfon hath been injured by cold
or moift air, or a high wind acting on the
lungs and outer furface of the body, that they
produce the fame effects by ftimulating too
much

much, as a sharp humour is known to do whilst it lies in the bowels or passes thence to the blood.

Q. Whether an abatement of the perspiration be of itself so certain and common a cause of fevers, as is thought, seeing the most dangerous oftentimes attack those suddenly who were in health but an hour or two before? Neither the quantity nor quality of that humour which might then be detained, can, with reason, be accused of such distressing effects in so small a time.

Whether, before a fever can come on, the fluids must necessarily have contracted some sort of acrimony or viscidity; and of course the solids be either too much braced or relaxed? I would be understood, that according to the common way of speaking, one or the other of those conditions must have existed, before the morbific cause began to act. Or are those of themselves sufficient to produce those effects? These two questions have been already answered.

Whence it it, in general, that the more severe the horror and the longer its continuance, so much the more dangerous is the succeeding disease? At first sight one would expect, that the stronger the rigor, the more powerful must be the cause that brought it

F   on,

on, and therefore the greater would be the difficulty to remove the diforder that follows. Yet, though this rule moſt commonly holds, we have an exception to it in the violent ſhiverings, which ſometimes precede tertian and quartan intermittents, the fits, in either caſe, often terminating in five or ſix hours; perhaps, becauſe the ſolids are lax rather than very ſpringy, and the blood not very compact, before thoſe ſorts of fevers can be formed at all; and therefore, the fibres relax ſooner after a good warmth is regained; the plentiful ſweat which follows putting an end to the fever for that time.

What is the reaſon why thoſe who commonly have cold feet, are in a particular manner liable to diforders of the hypogaſtrium and parts above? This want of heat muſt be owing to a lurking ſpaſm acting on the veſſels there, which prevents the free motion of the blood: ſo much of it therefore as cannot circulate freely in thoſe parts muſt be received by others, over and above what they ought to contain, if a regular diſtribution of the fluids took place. This ſurplus will be admitted wherever any local weakneſſes happen; ſo that the effects are extremely various. I never knew any perſon who was habitually liable to ſuch a coldneſs of the legs and feet, in whom an acid acrimony did not abound, at leaſt in the firſt paſſages, which by its irritation
tation

tation there, or having paffed thence to the blood, may indeed be the caufe of the partial fpafm we fpeak of.

Why are vomitings and purgings fo common in fevers during the Summer and Autumn ; and choleras, ferous or bilious diarrhœas alfo fo frequent in thofe feafons? Becaufe the folids are very lax at thofe times, and the blood much diffolved. The former therefore are exceedingly irritable and fufceptible of fpafms from flight caufes; and, feeing every unnatural conftriction of the outward parts repels the fluids in fome degree, they will now be immediately admitted by the veffels within ; in which, overfulneffes enfuing, their contents will be as readily allowed to efcape by an increafe of fome fecretion or other. The quantity of returning blood through the *vena portarum* to the liver being augmented by a fpafm, a profufe feparation of the bile follows; which, paffing into the duodenum, occafions vomitings, purgings, or both : for the gall is of fo pungent and faponaceous a nature, that it will ftimulate the ftomach and guts ftrongly to expel itfelf; and thofe difcharges will continue till the fpafm ceafes. However, this redundance of the bile is oftentimes prevented, by increafed fecretions made from other organs, as the ftomach, pancreas or inteftines : and we accordingly find, that the vomitings fometimes

F 2   are

are taftelefs, and the ftools pale and purely
lymphatic or ferous, fmelling more like frefh
blood than excrements.

May not the furprifingly great quantities of
urine, which hyfterical and hypochondriacal
perfons fometimes void immediately on the
attacks of fpafms, be eafily accounted for
from the anfwer to the preceding query? And
alfo, how thofe extraordinary fecretions in the
kidneys may divert greater evils, in the fame
manner, as if the overfulneffes of the veffels
within, were abated by vomitings and purgings ; as oftentimes happens to fuch fubjects,
when fpafms are brought on by fudden furprizes or frights ?

Does it not plainly appear from the foregoing theory, why, when the urine is clear
and limpid in fevers, from their firft attack,
or becomes fo after having been thick or high
coloured, a delirium, ftupor, convulfions, ftifling anxieties, or an obftinate watchfulnefs,
may with reafon be expected ? For fuch urine
fhows, that the fpafmodical conftrictions have
become more forcible, as well as general, or
that they were very ftrong from the beginning,
and amongft others reached the emulgent arteries themfelves. Therefore, as this fort of
urine is generally made in fmall quantities
during fevers, the fluids, for thofe and other
reafons, muft foon be dangeroufly collected
in

in the internal and fuperior parts if it be not prevented.

How comes it to pafs that either too great a flow or ftoppage of the lochia and menfes may happen from a fpafm? This is owing to the prefent local laxnefs or conftriction of the uterine and hypogaftric veffels, whilft the external ones are at the fame time more or lefs fpafmodically affected.

Whence is it, that at the approach of death fome perfons have better fpirits and are ftronger for a little while, the pulfe alfo being then fuller? This probably happens from an increafed and more general conftriction of the arteries, which urges the fluids in a fuller ftream towards the internal and fuperior parts; whereby a momentary greater fecretion of the animal fpirits is made, as their laft effort to maintain life. This is commonly called a lightening before death.

May not the painful ftiffnefs that is perceived in the joints and mufcular parts, more efpecially about the thighs and legs, before a fever, be more reafonably explained by fuppofing a fpafmodical conftriction of the arteries and other mufcular membranes, (perhaps of the fiefhy fibres of the mufcles themfelves) and the hinderance thence arifing to the free motion of the blood, (and it may be of the

F 3   animal

animal fpirits alfo,) rather than a too great
vifcidity of the fluids to allow a proper fecretion of the fpirits to be made, as fome authors
fuppofe? A plentiful fweat feldom fails to
abate thofe fymptoms, and fometimes removes
them at once; whereas, they ought indeed
to be increafed thereby, had they been owing
to too thick a confiftence in the blood.

Do not the palenefs and contraction of the
face and all the outward parts, together with
every other appearance in time of the horror,
clearly fhow that the fkin and the feveral parts
having any immediate connexion with it are
then ftrongly conftricted? And does not the
fmall hard pulfe make it equally plain, that
the coats of the arteries likewife fuffer in the
fame way; and that they continue thus difordered throughout the difeafe, fo long as the
pulfe is hard, (no matter whether it be fmall
and quick, or ftrong and flow,) as in comas,
apoplexies, and fome other diforders? And
feeing fo large a veffel as the carpal artery undergoes fo forcible a conftriction at the firft
onfet of a fever, may not the large trunks
and even the mufcular fibres of the heart itfelf fuffer likewife in the fame manner at
times, more efpecially when difeafes draw
near their mortal periods? This tightnefs in
the pulfe is imputed by fome to too great a
denfenefs of the blood: But this cannot be, for
the fame happens in fome degree during the
fevers

fevers of infants, anafarcous and other much relaxed or weakly people, in whom, furely, the fluids cannot with reafon be accufed of that fault.

Do not the great thirft that comes on before a horror attacks, and continues unquenchable in it,—the dry hard cruft likewife which prefently covers the tongue, and the drynefs of the mouth throat and noftrils, make it very plain, that the excreting ducts of the falivary and mucous glands, are then ftrongly contracted; whence, the thin parts of their humours exhaling, what remains is ftopped in the extremities of the ducts, there thickens, and together with the epithelium forms the above dry fkin? Some authors inform us, that this thick cruft arifes from a general diffipation of the thin parts of the blood, or I know not what feptic matter; but neither of thofe can be affigned for its caufe fo early as it oftentimes is feen. And this is alfo true, that the thicker and drier it is, fo much the greater commonly is the danger; that being a fign of general ftrong fpafms. But whenever thofe unnatural conftrictions are removed, this will foon difappear.

Are not the naufea, the diftention and throbbing about the epigaftrium, the pain in the loins, and the coftivenefs fo common in fevers owing to an overcharge in the internal

veffels,

veffels, and a fpafm that then acts on them, which prevents fuch an increafe of the fecretions there as might abate this fulnefs, rather than to any acrimony, (though now and then fomewhat of that fort takes place in the firft paffages); a fimple drynefs of the inteftines, which term has no precife meaning; or laftly, to the want of excrementitious matter in them, (though they frequently are much loaded therewith)? The pain in the loins may in part arife from the mefenteric nerves being over-ftretched by the furcharged bloodveffels, many of which are clofely furrounded by twigs of thofe nerves (1).

Are not the precipitate purgings which fometimes happen towards the end of fevers, moft commonly owing to an aggravation of the fpafm that acted before, whereby the fluids are ftill more repelled towards the internal parts? Becaufe the fpring of the veffels there was weakened by the continued overfulneffes they already fuffered, they will now yield the eafier to this additional weight of the fluids, and permit their contents to efcape: Though fuch purgings are commonly imputed to fome unknown acrimony or putrefaction of the humours

(1) *Vide* Vieuffen: Willis: Winflow: *paffim.*

Do not fudden furprizes and frights, act immediately as fpafms? The coldnefs, trembling, pale contracted countenance and fhrinking of all the outward parts; the palpitations of the heart, anxiety, quick-breathing, fobbing, weeping, fmall confufed frequent pulfe; the involuntary paffing of the urine and excrements, and rifing of the hair on-end plainly prove, that the external furface is ftrongly conftricted, and the fluids impetuoufly repelled thence by a fpafm, or fomewhat of this nature, call it what you will.

Whence are clammy fweats, and why do they prefage fo much danger, when the patient's ftrength is already fpent? And why do a palenefs of the face and fhrinking of the features at fuch times, fo certainly forebode death? Anfwer: Becaufe thofe are figns of very general and ftrong fpafmodic contractions, not only of the arteries, but of the fkin and other mufcular membranes likewife, whereby whatever was contained in the outward excreting canals is preffed out. The nofe is now pinched; and the lips, ears, temples and face are fhriveled, cold and pale, all the outward parts being fo forcibly conftringed that even the flefh itfelf feels ftiff and hard, as having loft its pliancy.

I may be thought tedious in mentioning thofe things: but the defign was to fhow,
how

how eafily many fymptoms which happen in
difeafes may be explained; and how a to-
lerable judgment may be formed, either of
the patient's recovery or death, by comparing
what hath been faid here, together with the
fecond and third chapters of this effay, in
which a fpafm was affigned for the immediate
caufe of acute diftempers; that a very general communication of this conftriction muft
happen, before they can prove mortal to
the fick, and alfo that it muft be removed,
ere the patient can recover. I have already
obferved that *Hippocrates* had fuch a notion
of fevers and their fymptoms. And fome mo-
derns reafoned well on the effects of fuch unnatural contractions; yet, in my opinion,
they did not alter the method of cure fo
much for the better, as might have been expected from their great abilities and long experience.

It hath been faid, that when the patient is
coftive, I loofen the belly at the beginning of
almoft every fever, and do not fuffer it to be
long bound afterwards in any one acute difeafe, provided the nature of the complaint
and condition of the fick will at all permit
his having two or three ftools daily; experience having convinced me, that the prefent
painful or dangerous fymptoms may be miti-
gated, and others prevented thereby, if proper care be taken in other refpects. In fhort,
I am

I am bold enough to fay, that whoever pur-
fues this method of emptying the abdominal
veffels fufficiently, and, in confequence thereof, thofe of the other internal and fuperior
parts, in the firft days of inflammatory and
common continued fevers, under the management above laid down, will remove them
fooner, and with lefs lofs of ftrength to the
patients, than by any other treatment hither-
to propofed. For, in few words, to recover,
thofe who are ill of fuch diforders, nothing
more feems wanting, than to abate whatever
overfulneffes may have happened within, from
an inverted circulation of the blood, caufed
by the fpafmodical conftrictions, that feem to
act chiefly and moft ftrongly on the external
veffels ; though thofe contractions muft alfo
be removed, otherwife the other end will not
be anfwered.

I frequently mentioned the neceffity for
laying hot bricks at a convenient diftance from
the feet and legs, in fevers ; and truly I have
found them of fo great fervice, that I think
they can fcarcely be difpenfed with in very
many acute and chronical complaints, more
efpecially during the winter, unlefs thofe parts
are inflamed. And I here declare, that they,
alone, have given more immediate relief in
hyfterical and hypochondriacal fpafms, than
could have been reaped from the whole *mate-
ria medica* of antifpafmodics without them ;

4

the

the fuffocations or convulfions oftentimes
abating prefently, without the affiftance of
any thing elfe (1). Putting the legs in warm
water, is frequently practifed on fuch and
other occafions: but the good effects thereof
generally ceafe with the immerfion ; for before thofe parts can be wiped dry, they. may
be again chilled, and the fpafm confequently
renewed. Befides, without great care the
patient's linen and bed clothes will be damped by the fteams of the hot water, which,
as things now are, muft have bad effects. It

(1) The juftly celebrated Baron *Van Swieten* quotes
from *Trillerus*, the cafe of a young man who had been
twice plentifully bled in a pleurify, and with advantage
as was thought. But on the fifth day the fymptoms
were fo violent, that the patient's life was in a manner
defpaired of. His feet being put in warm water, a vein
was opened in each, with a large orifice, but no blood
iffued, fo cold were thofe parts. Scalding water, as
would feem, was then poured on them, which made the
veins bleed freely, and with fo good effect, that the half
dead patient was not only relieved, but recovered.———
Vide *Van Swieten comment. in Boerhaave* Aph. Tom. III.
pag. 39.
This noble and excellent phyfician fpeaking here of
the advantages of bleeding in a pleurify, imputes the
recovery of this patient to the quantity of blood that
was then taken away ; though the two former large
bleedings did not prevent his growing worfe. And even
here it is more probable, that the lafting good effects
afcribed to that operation were more owing to the hot
water, which removed the fpafmodic conftriction that
acted on thofe parts. This inftance is exactly to the
point; and clearly fhows how neceffary it is to keep
up a good warmth in the feet and legs in fome acute
difeafes.

is ufual alfo, to wrap up the legs and feet in hot flannels ; but thofe foon cool and require fhifting often. Bottles filled with hot water have likewife been recommended; but I have known them break or leak at the corks in the bed, much to the detriment of the patient ; neither does the covering lie fo clofe as over the bricks, which may be had almoft every where, are foon made ready, and eafily changed when they cool. They alfo are much to be preferred to ftones, which are apt to fplit and fly in the fire; as I have feen happen even after they were put in the bed.

Thus have I given my opinion on this fubject ; which if founded in reafon, did not need the ornaments of eloquence, to recommend it to the ingenuous. But if the theory be not rational, or, what I reckon upon more, fhould not the practice ftand the teft, no flourifhes of ftile could protect them from that contempt which whimfical propofitions juftly deferve, whatever fpecious arguments introduce them to the public. Being confcious how little I can ferve mankind, I would by all means be careful how I propagated a doctrine, which, if wrong, would not only expofe me to cenfure, but might be deftructive in the higheft degree.

The newnefs of the practice in pleurifies, peripneumonies, and fome other acute difeafes,

eafes, may caufe it to appear in a fufpicious light to thofe, who do not comprehend the principles on which it is eftablifhed; or to thofe who are not accuftomed to reafon in the fame way on the immediate caufe of fevers, and the feveral fymptoms that happen in confequence thereof. But whatever may be their fentiments on this head, I hope the method propofed above will neither be condemned by implication nor prejudice, before it hath undergone fome fair trials in the way I have directed; when, provided the medicines operate properly, and due care be otherwife taken, I dare affirm, from the numberlefs inftances I have feen of its good effects, that it will feldom fail anfwering every purpofe we can poffibly defire from it. I do not pretend to the knowledge of any method or remedy which will remove any one diftemper with abfolute certainty, being very fenfible that the virtues of all medicines are only conditional or relative. Accordingly we now and then are mortified, by difappointments in the cure of thofe difeafes, which we vainly believed were beft underftood by us; owing, perhaps, to fomewhat of a hidden nature, a certain Θεῖον τὶ that fets bounds to our prefumption: But whether the difficulties arofe from my own overfights, fome peculiarities either in the conftitutions of individuals, or the diforders themfelves, which I could not difcover, fo it was, that I failed in fome cafes,

when

when I thought myfelf almoft fure of fuccefs.

*A Caution concerning Bleeding, and giving
Vomits,*

Scarcely any thing deferves more reprehen-
fion, than the rafh, undiftinguifhed, and repeated free ufe of the lancet; for however
neceffary it may be to take away blood on
particular occafions, beyond all doubt, this
is frequently done when it ought to be
omitted. Nor does any difeafe require fuch
a wafte of the vital fluids, as is too often
made; by which means the patient is left to
ftruggle with the double danger of the diftemper and extreme weaknefs.

Were it duly confidered, that the ftrongeft
perfons who eat and drink moderately, and
ufe neceffary exercifes, have no more blood
than is requifite in health; and that the plethora or overfulnefs fo much contended for,
at the beginning of acute and inflammatory
fevers, is for the moft part more imaginary
than real, and local rather than general; the
pretence for taking away fo much blood at
once, or by repeating the operation again and
again, would by no means appear to be founded in reafon. What then fhall we fay of thofe
who practife this on perfons who fall fhort of
fo vigorous a conftitution?

The

The arteries are naturally fo difpofed to contraction, from the ftrength of their circular fibres, that were it not for the refiftance made by our fluids they would fhut themfelves up to their axes. A certain quantity of blood muft therefore be neceffary to keep thofe veffels duly dilated, and excite a proper diaftole : for when too much of that fluid is loft, and the oppofition to their contraction is thus leffened, they will clofe themfelves in proportion; and the more, becaufe the force of the heart is weakened both by this great lofs of blood and the difeafe. We have a plain proof of this from the fainting that happens when the veffels have been fuddenly emptied by a large orifice ; for the quantity of returning blood to the heart is now fo diminifhed, that enough thereof does not arrive at that organ, to promote a proper diaftole that a forcible fyftole may follow. In this cafe the feet and hands become cold, and the countenance pale and contracted ; all the outward parts fhrink ; and clammy fweats appear as the arteries collapfe, for want of a fuitable quantity and due impulfe in the blood, to keep them fully diftended ; fo that the fkin, together with its contiguous veffels and membranes, plainly undergoes fpafmodical conftrictions at fuch times.

On fome occafions, we are directed to take away blood *ad animi deliquium :* But to fay
the

the leaft, this is a very uncertain rule; for
fuppofing bleeding to be neceffary, fome perfons may faint before a fufficient quantity hath
been loft, and others may bleed too much ere
they fall into a fyncope. But when a fainting
happens, the vein ought to be bound up immediately, however fmall the difcharge might
be, and the patient laid down if he had been
bled fitting up.

Repeated bleeding is alfo enjoined in fome
cafes, till the blood appears florid and free of
all fizinefs : But whatever fhow of reafon may
be in this, the injunction feems not much
better founded, than the common expreffion
*of taking away all the bad blood, and leaving
only what is good behind.* For it may be truly
afferted, that this point will not be attained,
if the fever continues, before the veffels have
been too much emptied, and the vital actions
depreffed in proportion.

The weaknefs brought on by an exceffive
lofs of blood is fudden and great; and its effects may be traced for a long time, by a palenefs of the face, trembling, chillinefs, and
want of heat in the extreme parts, great lan-
guidnefs of the vital and natural actions, and
a flow recovery. On the whole, though the
lofs of fome blood may be, occafionally, needful; or taking away a moderate quantity may
do little harm in fome cafes, the abufe of that

G   operation

operation is certainly very hurtful for the prefent, as well as the time to come. Eight or
ten ounces of blood are as much as any
man ought to lofe in moft diforders: for if
farther evacuation be neceffary, he that un-
derftands the foregoing theory cannot be at
a lofs as to the way it fhould be made; the
method above propofed having this great advantage, that though a patient may with fafe-
ty be much more emptied thereby than by
bleeding, this, however, is done by degrees,
and his ftrength is all the while fupported
with proper nourifhment. The judicious
Sydenham ferioufly laments the neceffity he
thought there was for bleeding fo often in
the rheumatifm, becaufe of the great weaknefs it brought on. He therefore changed
that practice, and, after the third bleeding,
depended on purging and anodynes, which
anfwered equally as to the difeafe, far better
furely for the patient: and the effects would
have been as good after bleeding only once,
or, perhaps, not at all, as much experience
hath convinced me; more efpecially had he
added fudorifics and alkalies to his other
medicines.

Vomits are thought by many to be, as it
were, indifpenfibly needful, more efpecially
in remitting and intermitting fevers: But, as
might be expected, thofe who make this a

rule

rule oftentimes do much mifchief; though
it is not charged, as it ought to be, to their
own temerity and want of judgment. All
the bad effects of this headlong fort of practice, cannot be recounted here : But let us
give a very common inftance in the cafes of
thofe who are ftupid and drowfy in fevers,
from too great a fulnefs in the veffels of the
brain, the belly being at the fame time bound.
—Now, though the fevers may abate, according to the ufual courfes of thofe difeafes,
yet a hard and too quick pulfe, head-ach,
ficknefs at the ftomach, and general diforder,
frequently continue as evidences, not only of a
lurking fpafm, but likewife of too great an
overcharge ftill remaining in the internal and
fuperior veffels. To give an emetic, therefore, under circumftances like thefe, can
fcarcely fail producing pernicious effects; even
convulfions, a delirium or ftupor fometimes
enfuing immediately, or at the next return
or increafe of the fever. How thofe things
fhould be, is fo evident as not to need any
explanation : Neverthelefs when a patient
difcharges much mucus, acid matter, green or
yellow bile, in the fever, we ought to promote the vomitings with warm water drank
plentifully till it is thrown up again clear and
taftelefs. Carduus and camomile tea are often
ufed for this purpofe, from a notion that they
ftrengthen the ftomach; but they lie fo little

a while

a while there, that any effect they may have
this way, does not deferve our regard at fuch
times: befides, the bitternefs of thofe infu-
·fions conceals other taftes, fo that we cannot
fo well know when all the offending matter is
difcharged, as if water alone were ufed.

ESSAY

ESSAY II.

Of the Crises of Common Continued and In-flammatory Fevers.

IT will plainly appear from the fore-going Essay how ill-qualified the author is for medical disputation, more especially, when the design was to oppose some notions, that have been long admitted for truths. But if there be reason to suspect that certain opinions have been adopted, which were liable to objection, it can have no bad consequences to examine them; seeing those things will still remain as before, if they be not disproved.

Various are the ways in which crises are said to be brought about: But, however the symptoms might be moderated in fevers by an hæmorrhage, or an increase of some excretion or

other,

other, I never saw an acute difeafe go clearly off before a general fweat came on. This fort of perfpiration, fometimes, is of fo rank a nature as to be offenfive to the fick themfelves, and their attendants alfo, owing, perhaps, to the great heat and rapid motion the fluids fuftained, whereby their faline and fulphure-ous' parts were not only multiplied, but ren-dered more volatile alfo; the urine and ftools, likewife, acquiring a more naufeous fcent from the fame caufes.

When this fweat is coming on, the pulfe becomes fuller and more foft, and the tongue moifter, which are figns that there is fome abatement of the fpafmodic condition of the folids, and that the hindrance to a more re-gular diftribution of the fluids is now giving way. And, as the perfpiration advances more, the pulfe ftill foftens, fills and becomes flower; whatever overfulneffes happened in the inter-nal and fuperior parts, together with the fe-veral fymptoms which depended thereon, abating alfo by degrees. For as the outward veffels relax yet more, and continue to dif-charge themfelves plentifully, thofe within, being eafed of their overload, by the humours now taking another courfe, recover their fyftole, and by little and little clear themfelves of the furcharges they fuffered; fo that, pro-vided things be but well managed now, a free circulation will foon enfue; or, in other words, the

the difeafe will ceafe, all the unnatural conftrictions being then abated.

Before the pulfe becomes fofter and more
full, it will be vain to look for the *fedimen-
tum album læve & æquale*, fo much wifhed for
in the urine during an acute difeafe; becaufe
the morbific matter is thought to be difcharged
that way. But this fort of fediment, among
others, is only a fign, that the veffels within
are recovering their natural diameters, whether they had been too much dilated or contracted before ; for more or lefs of fuch a
fettling, will always be feen in the urine of
people in health : And why it fhould be more
plentiful at the end of fevers, may be underftood from the greater laxnefs of the folids,
and attenuation of the fluids, of which both
the full foft pulfe and critical fweat are proofs.
Befides, from the vehemence of the circulation during a fever, more particles may be
rubbed off the coats of the veffels than in
health, and help to increafe this fediment.
Here I would be underftood to mean, what
happens at the termination of acute difeafes ;
for though the blood be ever fo fizy, when
the compreffing force of the veffels was ex-
erted with its greateft power, yet, if the patient be bled fome hours after the fever ceafed, the ferum will then be more in quantity
than it was during the diftemper, and the red
part lefs. Indeed, the palenefs and weaknefs

which

which are feen after every fever, plainly enough
fhow, that the folids are much relaxed and
the blood thinned.

It muft not be inferred from thefe loofe
hints, that I confider the infpection of the
urine as unneceffary ; on the contrary, I do
not know any of the internal fecretions from
which fo good a judgment may be formed
of the condition of the parts within during
fevers, as from that difcharge. The urine is
clear in acute difeafes with convulfions, ftupors, or delirium, (more efpecially if the latter be of the outrageous fort) in manias, apoplexies, epilepfies, during the horrors of all
fevers, as well as in hyfterical, hypochondriacal and afthmatic attacks, particularly when
the belly is bound in confequence of a fpafmodic conftriction of the mufcular fibres of
the inteftines, which prevents the fecretions
being made in them. But, when purgings
happen in fevers, the urine is, commonly,
thick or high coloured, and hath either a large
fediment or cloud ; for the urinary organs,
then partake of the fame laxnefs with the
other internal parts.

Limpid urine is juftly efteemed an unfavourable fign in fevers, for it fhews a very
extenfive ftrong fpafm. But the mifchief this
portends ought not to be imputed to the detention of any morbific matter that fhould
have

have pasfed that way, seeing the disease continued whilst the urine was very thick, but
to the spasm having communicated itself to a
greater number of vessels, and becoming
stronger also on those it possessed before,
whereby the fluids are repelled into fewer
vessels and of a smaller compass. In this case,
the most alarming consequences are to be
dreaded, from the overfulnesses and stagna-
tions that must soon ensue in the internal and
superior parts. And we accordingly see, that
either a delirium, convulsions, comatous or
lethargic stupors, mortal anxieties, &c. in-
fallibly succeed when such a change happens
in the urine, unless they can be prevented:
but this is seldom in our power, when patients
are much weakened.

Symptoms like those we have just now
mentioned, are said to happen *per metastasin*,
or from a translation of the morbific matter to
the brain or elsewhere. But that sort of reasoning explains nothing to me, who have
some doubts of such matter existing in the
fluids, more especially at the beginning of
common continued, inflammatory or even
intermitting fevers, when such symptoms are
frequently seen. However, I can easily comprehend how they may be brought on, by an
aggravation of the spasm; which, as it must
take place before a fever can attack, so it
must be removed ere the patient will recover,

or become very general before a difease can prove mortal, as already hath been faid.

When the urine is clear in fevers, it is commonly fecreted in fmall quantities: But hyfterical and fome other people, whofe nerves are very fenfible, oftentimes difcharge a very large quantity of urine immediately after the attack of a fpafm ; which, beyond all doubt, prevents ill confequences in another way, that otherwife would enfue when the fluids are thus precipitately and impetuoufly repelled towards the vifcera and brain. Again, when the belly is loofe in fevers, more efpecially during warm weather, the urine is generally high coloured, or full of fediment ; and fo long as it continues fo, there is not much danger either of ftupors or convulfions, unlefs the patients fit up too long when they are very weak. At the time therefore that the urine is clear in the firft days of fevers it is a rule with me to remove the ftricture of the veffels within, and abate whatever overful- neffes have happened there, by promoting loofe ftools and fweatings likewife, in order to obviate the danger that would quickly fol- low from the fluids being accumulated in the internal and fuperior veffels. This, in- deed, cannot be done when the patient's ftrength is already fpent by a lingering dif- eafe ; but it may and ought to be attempted at the beginning of common continued and
inflammatory

inflammatory fevers, which oftentimes prove
mortal on the third, fifth, or seventh day;
for in such cases, enough of fluids generally
remain to the last breath to support life, were
the veffels but in condition to circulate them.
They therefore who fall by diforders of that
fort truly die a violent death, as we have already faid. On the whole, though the ex-
amination of the urine be ufeful, as thence
both the approach and completing of the
crifis may be pretty well known, yet in my
opinion, it cannot be affirmed with any degree of propriety, that the body is fo commonly freed of the febrific caufe that way as
is thought, if it be at all, unlefs the diftemper arofe from gravel in the paffages : But we
do not fpeak of fuch complaints.

A crifis is faid to be brought about by
bleedings from the nofe, uterus and hæmorr-
hoidal veins ; and likewife by a diarrhœa, an
expectoration, &c. It cannot indeed be denied, that a well-timed fuitable difcharge
made in either of thofe ways will much abate
the painful fymptoms, by venting fome part
of the fluids that were too much collected in
the veffels within, and more fo in fome than
others, which being thus eafed of their overfulneffes will now contract themfelves with
greater power. And becaufe the fame effects extend to others, in fome degree, and
alfo to the heart and lungs, the blood muft
then

then be propelled with a greater impetus towards the conftricted veffels, and fo dilate them gradually. Yet, after all, the fweat we mentioned muft ftill enfue before a complete crifis can happen, by whatever natural or artificial difcharges the overload, which took place in the veffels within, might be leffened. Does not what we have faid of the humours being too much repelled from the outward to the internal parts at fuch times, plainly appear, among other things, from what happens to the ftrongeft man at the firft onfet of an inflammatory fever? For the pulfe is then fo fmall, and the patient likewife in all refpects fo exceedingly weak and helplefs, becaufe the vital actions are as it were overpowered, that they who know not the reafon, believe no blood ought to be taken away; though after that hath been done the fick perfon perceives himfelf ftronger, and the pulfe alfo becomes fuller.

Fevers are faid to ceafe when the morbific caufe happens to be thrown on the glandular or other parts: But, though it be eafy to conceive how obftructions of this fort may be formed, yet in what manner they prove truly critical, thofe muft declare to whom fuch inftances have occurred. For my own part, I always obferved, that fome degree of fever ftill remained till thofe tumours were either difcuffed or fuppurated; and even

then

then the patient fweated at laft before he recovered.

As to a critical diarrhœa, it is juft what we propofed to bring on early in the difeafe; and if the patient hath the good fortune to fweat, when that happens of its own accord, the fame falutary effects may probably enfue, as we find by promoting thofe difcharges at the beginning, in order to relieve the veffels within of the furcharges they fuftained, and perhaps clearing the firft paffages of fome fort of acrimony that might either bring on or continue the fpafm. But, whether this purging come on naturally, or was promoted by art, a general fweat muft neverthelefs follow, before the patient will be free of fever.

Whether fevers can be removed in cold climates without the fweat we mean, I will not difpute; but in *South-Carolina* they never have a crifis any other way: And how it can be otherwife elfewhere does not appear from the laws of mutual action between the folids and fluids, or according to the foregoing theory. The difcharge may not, indeed, be fo plentiful in cold countries as with us, in whom a laxer conftitution commonly prevails for the greateft part of the year.

My

My opinion of thofe things is confirmed, by what authors have faid of the different forts of crifes; as fuch increafed excretions could not have been made, unlefs the fluids were either too much collected in, or too plentifully determined towards the places whence they proceed. For it is plain, that had they circulated freely and equally throughout the fyftem as in health, nothing of that nature could have happened, becaufe the fecretions would not have exceeded any where; every part then receiving only its proper quantity of the fluids. Much lefs could a hæmorrhage break out, except the veffels had firft been fo over-ftretched that their coats were made to give way to the additional weight and impulfe of the blood. Neither could this have been if the progrefs of the fluids had not been intercepted, in fome degree, by fome change made in the condition of the arteries in other parts. It then feems clear, that fo much of the blood as could neither be received nor circulated by the veffels, which were fpafmodically affected, muft have been admitted by others, over and above what they ought to contain had a regular diftribution of it taken place, as we have already mentioned. An expectoration can no more happen in a peripneumony than from a diarrhœa, unlefs the fluids be firft præter-naturally accumulated in the lungs fo as to augment the fecretions there; and the extrava-

fated

fated humours, acting as a ftimulus, excite a cough to expel themfelves. Have we not evident proofs of too great diftention in' the veffels, at or near the place whence a future hæmorrhage or an augmented fecretion is to proceed, for fome time before this happens? Is not a bleeding from the nofe preceded by a fulnefs and flufhing of the face, throbbing of the carotid and temporal arteries, buzzing or ringing in the ears, and a heavinefs or pain in the eyes and forehead? How difficult is breathing, and what heat and oppreffion are perceived about the breaft in peripneumonies, and in the peripneumonic ftages of pleurifies, owing to the great weight of fluids, with which the lungs are charged. In the mortal ftages of all difeafes, how anxioufly do the patients tofs themfelves about, and gafp in the quickeft manner, becaufe of the ftifling from the fluids being moftly accumulated in the lungs and large veffels near the heart; the circulation being already at an end in the extreme parts. May it not be foretold that the menfes will foon flow, when a heat, throbbing, heavinefs and pain are perceived about the loins, hips and hypogaftrium? And, not to mention other things of the fame fort, when the circulation begins to recover itfelf in the external fmall arteries, the critical fweat may be announced by a more equal heat, a plumpnefs of all the outward parts, and foft-

nefs

nefs of the fkin, together with a fuller and lefs hard pulfe, a moifter tongue, and an abatement of the ficknefs, anxiety, &c. all plain proofs (of what hath been repeatedly alledged,) of a fpafmodic conftriction being the immediate caufe of fevers, which the feveral fymptoms we have laft recounted fhow is then giving away.

T H E   E N D.

www.ingramcontent.com/pod-product-compliance
Lightning Source LLC
Chambersburg PA
CBHW021944190326
41519CB00009B/1133